CW00515238

UK Nature Conservation No. 5

The ecology and conservation of European owls
Proceedings of a symposium held at
Edinburgh University

edited by

C.A. Galbraith, I.R. Taylor and S. Percival

assisted by S.M. Davies

Further information on JNCC publications can be obtained from Publications Branch,
Joint Nature Conservation Committee, Monkstone House, City Road,
Peterborough PE1 1JY

Typeset and Printed by The Charlesworth Group, Huddersfield, UK, 0484 517077

CONTENTS

Preface

During the last 20 years there has been a rapid increase in the number of scientific studies of owls in Europe. These have been motivated by a desire to understand the ecology of these fascinating and previously little known birds and by increasing numbers of conservation problems. As top predators, owls are particularly sensitive to changes in their environment. Across Europe several species, including the Barn Owl *Tyto alba*, Little Owl *Athene noctua* and Eagle Owl *Bubo bubo* have suffered serious decline.

After a number of preliminary discussions, centred mainly around these conservation problems, we decided it would be a good idea to try to bring together as many European researchers, and others interested in owls, to exchange ideas on methodology, and to discuss our current state of knowledge. And so this symposium came into being. It was organised by the Nature Conservancy Council, the British Trust for Ornithology and the University of Edinburgh and held in the Department of Forestry and Natural Resources of Edinburgh University.

This volume contains the papers and posters presented at the symposium and we hope they will lead to a greater appreciation and better conservation of these wonderful birds. All papers have been subject to independent refereeing to a standard expected in relevant scientific journals.

Colin Galbraith
Iain Taylor
Steve Percival

Acknowledgements

The programme and financial affairs were organised by Steve Percival, Iain Taylor organised matters in Edinburgh, and Colin Galbraith was charged with putting together the proceedings. The University of Edinburgh provided facilities and very generous financial support. The Joint Nature Conservation Committee funded the publication of the proceedings.

Many people helped at various stages and we are immensely grateful to them all. Don Smith and Derek Scott took us away from the graphs and figures and reminded us all of what owls look like by their breathtaking colour slides and verbal commentary, spiced with insight and humour.

The staff of the British Trust for Ornithology, especially Tracey Percival and Nicola Bayman helped with administration, as did staff and students of Edinburgh University, especially Laurel Hanna, Dominic McCafferty, Jon Massheder and Connie Fox.

We are particularly grateful to a large number of our academic colleagues who so generously gave their time to referee the papers.

A number of people within JNCC and the Nature Conservancy Council assisted with the publication of these proceedings. These included David Stroud, Lissie Wright, Flis Murat, John Holmes, John Bratton, Lucy Butler, Tracey Fisher, Maureen Symons, Janet Parr, Jan Avey, Mary Roberts, Pat Hall, Linda Porter, Jackie Bailey, Sheila Gorman and Janet Willmott.

Introduction

C.A. Galbraith

Conservation action in the UK requires statutory agencies and government to make rapid decisions, often without the benefit of instantly available specific scientific information. Symposia such as this held on the 'ecology and conservation of European owls' are therefore, extremely valuable in providing up-to-the-minute information on current research on such topics. They serve an essential purpose in communicating results directly to the conservation organisations, who are then able to act on the basis of quality science. The main aim of this volume is to help educate people interested in the conservation of owls throughout Europe and indeed the rest of the world.

Recent changes in populations of European owls, including several notable declines, have attracted the attention of many researchers and conservationists. As top predators in many ecosystems, owls are of particular value as 'indicators' of the health of these ecosystems and of the environment in general. There is an urgent need to develop an effective monitoring scheme for owl species in the UK and indeed throughout Europe. We have at present a series of counts of some species in some areas but this needs to be integrated and coordinated as part of an international programme to provide effective information for conservation management.

Whilst the monitoring of owl numbers is of value to their conservation, it is also important to understand their current pattern of distribution and the underlying ecological factors affecting them. Detailed studies of the birds' feeding and other ecological requirements, and the factors limiting their populations all provide essential information. In addition studies of the genetic structure of populations, rarely studied in wild populations, can give insight to the species' ecology, often with important implications for conservation practice.

Conservation practice and management have been mentioned several times but what do they actually mean? Many organisations are currently contributing to owl conservation throughout Europe. This action ranges from the creation of suitable habitat for a range of species to nest box schemes and advice to landowners, farmers and governments. There is a clear need however, for these actions to be coordinated to maximize the benefits for conservation.

Take for example the Barn Owl *Tyto alba*. There is now more effort directed to the conservation of this species than any other owl in the UK, but is this action arresting the population's decline? We must improve the impact of conservation measures and attempt to introduce interested people to help in the most effective way. There has been a dramatic increase in concern about the conservation of the Barn Owl and many people have become involved in the release of captive birds into the wild. Whilst this is in many ways an understandable progression, it actually often leads to the early death of the birds released and may damage remaining wild populations. Clearly such reintroductions should conform to the generally accepted guidelines produced by the IUCN.

The papers in this volume will help improve our understanding of the conservation needs of owls and hopefully therefore of the actions that need to be developed to ensure their survival across their traditional range.

Population dynamics of Fennoscandian owls in relation to wintering conditions and between-year fluctuations of food

E. Korpimäki

Korpimäki, E. 1992. Population dynamics of Fennoscandian owls in relation to wintering conditions and between-year fluctuations of food. *In: The ecology and conservation of European owls*, ed. by C.A. Galbraith, I.R. Taylor, and S. Percival, 1-10. Peterborough, Joint Nature Conservation Committee. (UK Nature Conservation, No. 5.)

1. Small mammal populations are strongly cyclic in northern Fennoscandia, showing high peaks at intervals of four to five years, whereas the populations are more or less non-cyclic in southern Fennoscandia, showing only seasonal changes with high numbers in autumn and low numbers in spring. In the transition zone, as in southern and western Finland, they are weakly cyclic and only *Microtus* spp. exhibit marked between-year fluctuations. The depth of snow cover and the length of snowy period also increase northwards in Fennoscandia.

2. Owls can adopt two basic strategies when living in different parts of Fennoscandia. Resident species are generalist predators, show short-distance breeding dispersal, are long-lived, and have long-lasting pair-bonds and small clutches. Nomadic or migratory species specialise on small mammals, shift their nesting sites nearly every year, are short-lived, and have yearly pair-bonds and large clutches.

3. Among Fennoscandian owls, resident generalists are the Eagle Owl, Ural Owl, Tawny Owl and probably also the Pygmy Owl. Nomadic or migratory specialists are the Snowy Owl and Short-eared Owl, and perhaps also the Great Grey Owl, Hawk Owl and Long-eared Owl. Tengmalm's Owl shows an intermediate strategy.

4. The population dynamics, breeding dispersal and food of Ural, Short-eared, Long-eared and Tengmalm's Owls are described based on information in the literature.

5. The hunting mode (quartering v. sit-and-wait technique), diet choice (specialised v. generalised) and nest-site (open nest v. hole-nest) seem to be the three most important factors affecting inter- and intra-specific differences in seasonal migration patterns and tactics of breeding dispersal of Fennoscandian owls.

Erkki Korpimäki, Department of Zoology, University of Oulu, Linnanmaa, SF-90570, Uulu, Finland

Present address: Kp. 4, SF-62200 Kauhava, Finland

Introduction

Fennoscandia comprises Norway, Sweden, Finland and north western parts of the USSR (Kola peninsula and Karelia). For practical reasons the area within the former USSR has been left out of this paper.

Ten owl species (the Eagle Owl *Bubo bubo*, Snowy Owl *Nyctea scandiaca*, Hawk Owl *Surnia ulula*, Pygmy Owl *Glaucidium passerinum*, Tawny Owl *Strix aluco*, Great Grey Owl *Strix nebulosa*, Ural Owl *Strix uralensis*, Long-eared Owl *Asio otus*, Short-eared Owl *Asio flammeus* and Tengmalm's Owl *Aegolius funereus*) breed in Fennoscandia (Mikkola 1983). In addition, the Barn Owl *Tyto alba* has bred in southern Sweden, but its population is now nearly extinct (Holmgren 1983). In this paper, therefore, no further reference is made to the Barn Owl.

The primary prey source of these ten owl species are small rodents such as voles, mice and lemmings.

Alternative prey items include shrews, hares and birds (Korpimäki 1981, 1988a; Nilsson 1981; Mikkola 1983; Korpimäki & Sulkava 1987). The availability of these prey items as a potential food supply for owls is influenced by the characteristic winter patterns of the Fennoscandian area. Even a shallow snow cover is known to substantially reduce the availability of small mammals as a potential source of food (Korpimäki 1981, 1986a; Sonerud 1986).

Large regional differences in the snow conditions occur within Fennoscandia. For example, in southern Sweden the mean maximum snow depth is between 5 and 20 cm for a period of two to three months, whereas in northern Sweden the respective figures are 100 to 120 cm for a period of three to five months. In eastern and northern Finland, the depth and duration of snow cover varies again and is typically 60 to 80 cm for a period of six to seven months (Solantie 1975, 1977). The densities of small mammals and small game also show marked between year

1

variations in Fennoscandia (Angelstam, Lindstrom & Widen 1984, 1985; Hansson & Henttonen 1985; Linden 1988; Korpimäki & Norrdahl 1989a). These variations are called short-term population fluctuations and occur on a three to four year cycle.

This paper focuses on the population dynamics of owls in relation to both the wintering conditions and between year fluctuations in prey item numbers. Data are also presented for the Kestrel *Falco tinnunculus* which is also dependent on small rodents as its primary source of prey. Most of the data presented are from the author's studies on Tengmalm's Owl, Short-eared Owl, Long-eared Owl and Kestrel. These studies were carried out in the Kauhava region (*c* 63°N, 23°E) in the province of South Ostrobothnia, western Finland. Details of these studies are described in Korpimäki 1981, 1984, 1985, 1986a, 1986b, 1986c, 1987a, 1987b, 1988a, 1988b, 1988c and Korpimäki & Norrdahl 1989b. Between year variation in vole numbers is relatively straight forward to estimate using snap-trappings, although reliable density indices are laborious to obtain. The effects of fluctuating food supply on owl population dynamics can therefore be easily studied.

Geographical trends in the population fluctuations of small mammals and small game in central and western Europe and in Fennoscandia are presented, and the strategies of Fennoscandian owls in relation to prey availability and wintering conditions described. Examples of various dispersal strategies that owls adopt when living in different parts of Fennoscandia are given. Finally, a hypothesis to explain which factors govern the migration patterns and dispersal tactics of Fennoscandian owls is developed.

Geographical trends in the population fluctuations of owl prey items

Recent advances in our knowledge of geographical trends in the population fluctuations of small mammals in Europe have been made. Hansson & Henttonen (1985) analysed gradients of density variations of small rodents in Europe. They illustrated that 'cyclicity indices', measurements of the degree of between-year population fluctuations, for Field Voles *Microtus agrestis* and Bank Voles *Clethrionomys glareolus* were positively related to latitude. In Fennoscandia these cyclicity indices are better correlated with the maximum snow depth and the duration of snow cover.

Hansson & Henttonen (1985, 1988) divided Fennoscandia into three regions according to the

density variations of microtines:

1. a southern zone (from 55° to *c.* 59° N);
2. a transition zone (*c.* 59° to 61° N);
3. a northern zone (from 61° northwards).

Korpimäki (1986c, 1986d) showed later that the transition zone in fact extended from the southern coast of Finland to 65° N in the western part of the country, where the snow cover is shallow (for a tentative distribution of these zones see Hansson & Henttonen 1988).

The vole populations in the southern zone are more or less non-cyclic (Hansson 1984). Instead of pronounced between-year fluctuations a seasonal pattern can be detected with high numbers in autumn after the reproductive season, and low numbers during spring, prior to the following reproductive season (Erlinge 1987; Dijkstra *et al.* 1988).

In the northern zone vole populations are strongly cyclic and are characterised by high peaks at intervals of four to five years. In addition, the northern vole cycles are also characterised by steep population declines in the summer and very low densities during the crash phase. The crash phase lasts between one and two years (Henttonen *et al.* 1987) and is synchronous over large areas (Kalela 1962). The crash phase of co-existing shrew populations is also synchronous (Hansson 1984; Henttonen 1985; Sonerud 1988).

The microtine populations in the transition zone are weakly cyclic. Only *Microtus* species (the Field Vole and Common Vole *Microtus arvalis*) show marked year to year fluctuations with peak numbers attained at intervals of three to four years. In contrast Bank Voles do not usually exhibit large between-year fluctuations although seasonal patterns can be detected with low numbers during the spring and high numbers in autumn. The population dynamics of shrews tend to parallel those of the Bank Vole (Korpimäki 1986d).

Thus the pattern of fluctuations in small mammal populations in the transition zone differs from that in the northern zone. Compared to the northern zone, summer declines are unusual in the transition zone, peaks are lower, the crash phase is not necessarily synchronous in large areas and in all of the small populations the densities of the low phase are higher and the length of the crash phase shorter (Korpimäki 1981, 1986d; Korpimäki & Norrdahl 1989a).

Angelstam, Lindstrom & Widen (1984, 1985) and Linden (1988) also report that populations of small game, such as hare and grouse, fluctuate synchronously with vole populations in central and northern Fennoscandia. The cyclicity and synchrony are most marked in the north (i.e. in the northern zone of microtine fluctuations), but decrease southwards (Angelstam, Lindstrom & Widen 1985). Small game populations, like small mammal populations, are relatively stable between years in southern Fennoscandia.

Geographical trends in snow conditions and prey population fluctuations – the effect on owls

The timing and availability of small mammals as a prey source for Fennoscandian owls differs between zones. In the southern zone, vole populations are at a seasonal low phase during the early part of the owls' breeding season. However, during the independence period when the mortality rate of owls tends to be high (Southern 1970; Exo & Hennes 1980; Bairlein 1985) the availability of prey items coincides with the seasonal peak in vole numbers and is therefore high. Winters are also mild in this southern zone with only a shallow snow cover which again results in increased availability of small mammals. On the other hand inter- and intra-specific competition for small mammals is likely to be high in this zone. For example, Erlinge *et al.* (1983) showed that in southern Sweden voles were heavily predated by both mammalian and avian predators.

In the northern zone, vole supply fluctuates in a predictable cyclic pattern and plentiful prey sources are available for one or two breeding seasons during each microtine cycle. However, in this zone two other problems can occur. Firstly, whilst vole cycles may be predictable in time, they are not as predictable in space. Owls must therefore be able to locate a nesting territory which supports a sufficient vole density as well as an adequate number of nesting sites. Secondly, the owls must have a strategy for surviving crash phases of the microtine cycle, especially since the numbers of alternative prey, shrews and small game, will be in their low phases at the same time. Crashes of the vole populations may happen unpredictably during the course of the owl breeding season resulting in a marked reduction in reproductive success. Additionally, deep and prolonged snow cover during the winter will decrease the availability of voles at the surface. Vole densities below the snow layer may, however, still be high. Thus, the owls may adopt a survival strategy of moving into areas with little snow cover for the winter period.

Strategies of Fennoscandian owls in relation to wintering conditions and population fluctuations of their food supply

There are two basic strategies that owls can adopt when living in an environment which varies seasonally and where food supplies fluctuate between years (Table 1). Owls can either be generalists or specialists.

Resident species are generalists, shifting to alternative prey to survive 'lean' periods. Dietary shifts are possible due to adaptable hunting techniques which enable owls to take a range of prey types, such as small mammals and birds. Resident generalists remain on the same territory year after year or move only short distances between successive breeding seasons. They are normally long-lived, monogamous and strongly territorial. Their pair-bonds last for several years, even for a lifetime. Resident generalists do not normally breed as yearlings, and their clutches are usually small.

Nomadic or migratory species are specialists, feeding almost entirely on small mammals (Table 1). Their hunting technique is well adapted to capture preferred prey items, but they are poor at adapting to hunt alternative prey, such as birds. Specialists shift their nesting sites nearly every year in relation to the phase of the vole population cycle although they sometimes remain on the same territory for two breeding seasons during the increase and peak phases of the vole cycle. Nomadic specialists usually overwinter within Fennoscandia, whereas most migratory specialists move to central and western Europe in autumn and back to Fennoscandia in spring. Long distance breeding dispersal is adaptive, because vole populations fluctuate asynchronously in different parts of Fennoscandia. The likelihood of finding a breeding site with a high vole supply is therefore quite high. Other life history traits associated with this strategy are small body size, short lifespan, pair-bonds lasting for only one breeding season or a part of it, weak territoriality, and polygyny and polyandry. In addition, nomadic and migratory specialists usually breed as yearlings and their clutches are large.

Considering geographical differences in the snow conditions and population fluctuations of food, it can be assumed that the availability of alternative prey decreases northwards in Fennoscandia. The resident generalists are therefore best adapted to live in the southern zone whilst nomadic specialists are better adapted for the northern zone.

In Table 2, Fennoscandian owls are classified according to their tactic of natal dispersal. Detailed

Table 1. Tentative list of life-history traits associated with the two basic strategies of Fennoscandian owls. Sources within parentheses.

Life traits	Resident generalists	Nomadic specialists
Body size (1)	large	small
Breeding dispersal (2)	short-distance	long-distance
Life-span (3)	long	short
Pair-bond (4)	life-long	yearly
Mating system (5)	monogamy	polygamy
Territoriality (6)	strong	weak
Age of first breeding (7)	more than 1 year	1 year
Clutch size (8)	small	large
Diet choice (9)	generalized	specialised

Sources:
(1) Mikkola (1983), Korpimäki (1986e), (2) Saurola (1987), Wallin (1988), Korpimäki, Lagerstrom & Saurola (1987), Village (1987), (3) Korpimäki (1986e), Saurola (1987, 1989), Pietiainen (1989), Wallin (1988), (4) Saurola (1987), Korpimäki (1981, 1988c), (5) Korpimäki (1988c, 1989), (6) Erlinge *et al.* (1982), Mikkola (1983), (7) Korpimäki (1987e), Pietiainen (1988), (8) Mikkola (1983), Korpimäki (1986e), (9) Mikkola (1983), Korpimäki (1986e).

Table 2. Tentative classification of Fennoscandian owls according to their tactic of breeding dispersal. Sources within parentheses

Resident generalists		Nomadic or migratory specialists
Eagle Owl (1)	Tengmalm's Owl (2)	Snowy Owl (3)
Ural Owl (4)		Great Grey Owl ? (5)
Tawny Owl (6)		Short-eared Owl (7)
Pygmy Owl ? (8)		Long-eared Owl ? (9)
		Hawk Owl ? (10)

Sources:
(1) Mikkola (1983), (2) Korpimäki (1981, 1986c), Korpimäki, Lagerstrom & Saurola (1987), (3) Andersson (1980), Mikkola (1983), (4) Korpimäki (1986e), Saurola (1987, 1989), Pietiainen (1988), (5) Mikkola (1983), Korpimäki (1986e), (6) Korpimäki (1986e), Saurola (1987), Wallin (1988), (7) Andersson (1980), Village (1987), this paper, (8) Mikkola (1983), (9) Hagen (1965), this paper, and (10) Mikkola (1983).

studies only exist for ringed populations of Ural, Tawny and Tengmalm's Owls. Therefore for many of the species classification is tentative. For species marked by queries the classification is more or less a guess. Resident generalists are Eagle Owl, Ural Owl, Tawny Owl and probably also the Pygmy Owl. Nomadic or migratory specialists are Snowy Owl and Short-eared Owl, and perhaps also Great Grey Owl, Hawk Owl and Long-eared Owl. Short-eared Owls and most Long-eared Owls are migratory in central and northern Fennoscandia, but they seem to shift their nest-sites over long distances between successive years (Saurola 1983). Tengmalm's Owl is an intermediate species, as in central Fennoscandia most females are nomadic but most males are resident (see below).

Ural Owl: a resident generalist

The Ural Owl is probably the best studied resident generalist species among Fennoscandian owls. Its original nest-sites were hollow tree stumps and old stick nests, but nowadays most pairs nest in boxes (Mikkola 1983). Pietiainen (1989) studied how the proportion of breeding pairs varied between 1977 and 1988 in a population in southern Finland. The yearly number of nest-boxes occupied in the study area remained quite stable (60–80), but the number of breeding pairs varied markedly between years. Only 10 to 30% of pairs bred in poor vole years, whereas this proportion was 70 to 90% in good vole years. The data indicated that the shortage of food was the most important reason for non-breeding, although the possibility that in some cases it was associated with absence of a mate cannot be ruled out.

Saurola (1987) analysed Finnish ringing recoveries of the Ural Owl. According to these long term data, all males and 96% of females move less than 6 km between successive breeding seasons. This indicates a high degree of nest site fidelity by Ural Owls.

Two *Microtus* voles and the Bank Vole and Water Vole *Arvicola terrestris* together made up 64% of the prey items recorded in a population of breeding Ural

Owls in western Finland during 1973–85 (Korpimäki & Sulkava 1987). The yearly proportion of *Microtus* voles varied from 10 to 60% and that of Bank Voles from 5 to 25%. Both proportions were positively correlated with the spring trap indices of these voles. The diet choice increased in low vole years with shrews, young hares, birds and frogs forming the most important alternative prey source.

Short-eared and Long-eared Owls: migratory specialists

Population fluctuations, breeding performance and diet of Short-eared and Long-eared Owls were studied during 1977–87 at Alajoki farmland (Korpimäki 1984, 1987a; Korpimäki & Norrdahl 1989c). The study area covers 47 km^2 and primarily comprises flat cultivated fields (77% of area). Short-eared Owls breed on the ground and Long-eared Owls on stick nests, but some pairs also accept nest-boxes. In addition to these owls, the Kestrel is an important avian predator of small mammals at Alajoki.

The yearly number of breeding Short-eared and Long-eared owls varied widely between years during 1977–87 at Alajoki (Figure 1). The range in the number of Short-eared Owl pairs was 0–49, in Long-eared Owl pairs 0–19 and in Kestrel pairs 2–46. Although the number of non-breeding pairs was unknown, the populations of these three species seemed to contain very few non-breeders.

Two *Microtus* species, the Bank Vole and Common Shrew *Sorex araneus* are the most frequent small mammals at Alajoki. Spring and autumn densities in the four most important habitats were estimated by snap-trappings (Korpimäki 1984, 1987a; Korpimäki & Norrdahl 1989b, 1989c). The spring number of *Microtus* spp. exhibited large year-to-year variations with peak numbers reached in 1977, 1982 and between 1985 and 1986 (Figure 1). The number of Common Shrews peaked in 1982 and 1984, but in other springs it remained quite stable, as did the number of Bank Voles during the whole study period. The number of pairs of birds of prey fluctuated in close accordance with the number of *Microtus* voles (Figure 1) (Spearman rank correlation: r = 0.87 for the Short-eared Owl; r = 0.83 for the Long-eared Owl; and r = 0.78 for the Kestrel; *P* < 0.01 in each case), but not with the number of Bank Voles and Common Shrews. It is notable that the number of breeding birds of prey usually follows the population fluctuations of *Microtus* voles without time lag.

No data exist for nest-site fidelity of Short-eared and Long-eared Owls breeding at Alajoki. However, a

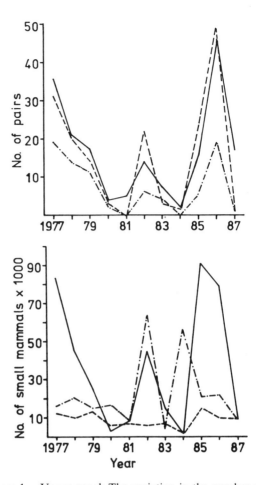

Figure 1. Upper panel: The variation in the number of Kestrel *Falco tinnunculus* (solid line), Short-eared Owl *Asio flammeus* (dashed line) and Long-eared Owl *Asio otus* pairs breeding at Alajoki during 1977–87.
Lower panel: The variation in the estimated total number of Field voles *Microtus agrestis* (solid line), Bank voles *Clethrionomys glareolus* (dashed line) and Common Shrews *Sorex araneus* at Alajoki during 1977–87. Data from Korpimäki (1984) and Korpimäki & Norrdahl (unpublished).

total of 191 female Kestrels and 181 males were trapped and ringed at their nests during 1983–88. Turnover of females was 94% and that of males 82% (Korpimäki 1988c). As the yearly survival rate of adult Kestrels is 50–60% (Cave 1968, Wallen *et al.* 1983), the data indicate that most females and males disperse widely. As population fluctuations of Short-eared and Long-eared Owls at Alajoki seemed to be more pronounced than those of Kestrels, their breeding dispersal may be even greater than that of Kestrels. Two Short-eared Owls wing tagged during a study in southern Scotland, in which a total of 21 Short-eared Owls were tagged, were later reported to be breeding more than 400 km away (Village 1987). This indicates long distance breeding dispersal in at least one European population. It can therefore be suggested that changes in the Short-eared and Long-eared Owl populations, breeding at Alajoki, are the

5

result of emigration and immigration, as the owls track the vole populations both inside and, to a lesser extent, outside Alajoki (Korpimäki 1985).

Microtus voles form the staple food of the Short-eared and Long-eared Owl diet at Alajoki. Data from 1977–87 indicated that these voles formed 58% of the prey weight of breeding Short-eared Owls and 72% of that of Long-eared Owls (Table 3). The proportion of *Microtus* voles increased to 90% in the diet of Short-eared Owls and to 80% in that of Long-eared Owls in good vole years. As vole numbers declined during poor vole years at Alajoki the proportion of voles in the diet of breeding Short-eared and Long-eared Owls decreased to 4–30% of the total prey weight. Bank Voles, Water Voles, shrews, mice and small birds served as the most important alternative prey (Korpimäki 1987a and unpublished).

Short-eared and Long-eared Owls breeding at Alajoki behave as migratory vole specialists in the Fennoscandian transition zone. Long-eared Owls are also field vole specialists in southern Sweden (Erlinge *et al.* 1983), but their degree of nest-site fidelity seems to be higher in central and western Europe than in Fennoscandia (Wijnandts 1984).

Tengmalm's Owl: an intermediate species

Tengmalm's Owl is a hole-nester which readily accepts nest-boxes. Studies of nest-box populations are popular in Fennoscandia and in Germany. There are at least 30 populations where the number of breeding pairs have been tracked using similar methods over a period of at least four successive years: six in Germany, one in Norway, four in

Table 3. The percentage prey weight of *Microtus* voles in the diet of Short-eared and Long-eared Owls breeding at Alajoki, western Finland, during 1977–87. Data from Korpimäki (1986b, 1987a and unpublished).

Year	Short-eared Owl	Long-eared Owl
1977	90.6	81.9
1978	71.1	76.2
1979	54.3	62.5
1980	3.8	0.0
1981	–	–
1982	41.5	69.1
1983	17.2	33.9
1984	24.7	–
1985	67.2	71.6
1986	69.9	79.4
1987	29.1	32.2
Total	**57.6**	**72.3**
No. of prey items	3626	3338

Sweden and nineteen in different parts of Finland (Korpimäki 1986c).

The amplitude of between-year fluctuations in the number of breeding pairs by calculating a coefficient of variation ('cyclicity index') for each population was measured. The correlations between cyclicity indices and latitude or longitude were significantly positive in Europe (Korpimäki 1986c). However, within Fennoscandia cyclicity indices were positively correlated with the mean yearly period of snow cover and the mean maximum snow depth, as opposed to geographical location. The amplitude of owl fluctuations in Fennoscandia therefore is more dependent on the amount of snow than on geographical location, especially on the southern and western coast of Finland where the snow cover remains shallow towards inland areas. Small owl population fluctuations occur in these northern areas which impairs the correlation between latitude and cyclicity indices in Fennoscandia.

Based on the data on population fluctuations of Tengmalm's Owl, Europe can be divided into three sub-areas (Korpimäki 1986c). The breeding populations are most stable in central Europe, then in southern and western Finland, and least stable in Norway, in central and northern Sweden, and in eastern and northern Finland.

During four successive breeding seasons Tengmalm's Owls were trapped in six populations in Europe (Korpimäki 1986c). Assuming a similar trapping efficiency in these areas, the proportion of owls later retrapped within the area can be assumed to reflect the degree of site-fidelity. In those populations where two sexes were trapped, the proportion of retrapped males was markedly higher than that of females. The proportion of both sexes later retrapped within the study area was significantly larger in Germany than in western Finland. On the other hand, a pair of Tengmalm's Owls from western Finland were more frequently found within the study area than conspecifics in Umea, northern Sweden. Although several sources of error may bias these comparisons, they do illustrate that the majority of females and males are resident in central Europe. Most females in western Finland are nomadic but males typically resident. The degree of nomadism of females and males in northern Sweden is higher than that in western Finland.

Data on both diet and population fluctuations were available for eleven Tengmalm's Owl populations (Korpimäki 1986c). Diet breadth correlated negatively with cyclicity indices of breeding populations, but no significant relationship was found between diet breadth and latitude. Tengmalm's

Owls from highly cyclic populations feed almost exclusively on *Microtus* and *Clethrionomys* voles, whereas stable populations of Tengmalm's Owl occur where there is greater availability of alternative prey such as shrews, small birds and mice.

In summary, one can say that the Tengmalm's Owl is a resident generalist in central Europe, but is nomadic and a vole specialist in northern Fennoscandia. In the transition zone, as in western Finland, Tengmalm's Owls seem to adopt an intermediate strategy with males typically resident and females nomadic. It is probable that other European owls, for which small mammals form the major part of the diet change their habits from south to north. However there is no existing data to support these changes.

Thus population fluctuations of voles affect the breeding density, degree of site fidelity and composition of the diet of Tengmalm's Owls in addition to influencing several other traits of population dynamics and breeding performance. Large between-year variations in the timing of breeding (Korpimäki 1981, 1987c), clutch size and number of fledglings produced depend on the vole supply (Korpimäki 1981, 1987d). Distances of breeding and natal dispersal tend to be shorter under good food conditions than under poor ones (Lofgren, Hornfeldt & Carlsson 1986; Sonerud, Solheim & Prestrud 1988; Korpimäki & Lagerstrom 1988). Polygyny and polyandry only occur in good vole years (Carlsson, Hornfeldt & Lofgren 1987; Korpimäki 1988c, 1989). The age composition of breeding populations changes in relation to vole numbers with a high proportion of juveniles or novice breeders in good years and a low proportion in poor years (Korpimäki 1988d). It is interesting that survival of fledglings seems to vary depending on the phase of the vole cycle (Korpimäki & Lagerstrom 1988). Fledglings which are raised during the increasing phase of a vole cycle survive the first winter twice as well as those raised in other phases of the vole cycle. An abundant food supply during the post-fledging and independence periods was the most likely reason for the high survival in the increase phase. As vole cycles are fairly predictable in time, a selective advantage can be expected for those owls that invest most in reproduction during the increase phase of the vole cycle, the contribution to the future gene pool of the population being highest at that time.

Does the availability of nest-sites or food govern the dispersal tactics of Fennoscandian owls?

Two hypotheses have been put forward to explain which factors govern the migration patterns and dispersal tactics of Fennoscandian owls. Firstly, Lundberg (1979) suggested that access to a nest-site is the most important factor responsible for shaping migration patterns (see also Wardhaugh 1984 for British owls). Secondly, Sonerud (1986) showed that during the snowy season small mammals were more available to hunting birds of prey in closed forests than in open, clear-cut areas. He assumed that interspecific differences in the ability to shift between hunting in open country and woodland habitats, and to capture small mammals protected by snow, governed seasonal movement patterns.

According to the first hypothesis, residency of the Ural Owl is an adaptation to the scarcity of nest-sites and is facilitated by a generalised diet (Lundberg 1979). Migration to snow free areas and long distance breeding dispersal of Long-eared Owls are possible as their preferred nest sites are more abundant than natural cavities. Migration is also an adaptive strategy as the species are specialised feeders on small mammals (Lundberg 1979). In hole-nesting Tengmalm's Owls that feed mainly on small mammals there is a conflict between residency and nomadism. Thus, Lundberg (1979) suggested that females are nomadic and males resident when small mammals crash. Extensive field data published later confirmed this suggestion (see above).

According to the second hypothesis, Ural and Tengmalm's Owls remain in areas with a high snow cover, locating small mammals by sound using an inexpensive sit-and-wait technique which is suitable for hunting in forests. Residency allows them to occupy energetically better nest sites, such as natural cavities (Sonerud 1986). Long-eared Owls do not normally arrive at central and northern Fennoscandia before the open ground is partially snow-free as their energy-expensive quartering technique for hunting is not suitable for a deep snow cover. Long-eared Owls therefore have to accept poor nest sites, such as old stick nests, as the resident owls have occupied the better nest sites in natural cavities (Sonerud 1986).

In my opinion, neither of these two hypotheses can alone explain inter- and intra-specific differences in migration and dispersal of Fennoscandian owls described here. As young owls have to overwinter at least once before their first breeding, the first year survival is determined more by the dispersal tactic than by the choice of nest site. Thus specialists which are unable to locate vole prey items that are protected by snow, or which hunt in forests due to their hunting technique, must migrate in the autumn. When they return in spring, they have to lay eggs on vacant nest sites. Sonerud's (1986) hypothesis may therefore explain why Short-eared and Long-eared

Owls are migratory in Fennoscandia and breed on open nests. This hypothesis, however, cannot explain why they shift their nesting territories over long distances and why the remaining Fennoscandian owl species are either nomadic, intermediate, or resident (Table 2). One should add at least two other factors, these being the choice of diet and nest site, to the hypothesis. Of these, the specialisation on small mammals seems to explain why Short-eared and Long-eared Owls have to search for nesting areas with a good vole supply, thus showing long-distance breeding dispersal.

The remaining eight species in Table 2 use an energetically cheap sit-and-wait hunting mode and can locate prey protected by snow (Sonerud 1986). Thus, Sonerud's (1986) hypothesis does not explain why they show different breeding dispersal.

Of the resident species, Eagle Owl breed on the ground and Ural, Tawny and Pygmy Owls in tree-holes. The Eagle Owl is large and can quite easily shift to alternative prey when small rodents are protected by snow or when their populations crash (Sulkava 1966; Willgohs 1974). In contrast, Eagle Owls, due to their long wings, cannot usually hunt in closed forest stands. Other aforementioned resident species are also generalists in their diet choice. The ability to shift to alternative prey during low phases in the vole cycle and during periods with snow cover is the most important factor promoting residency among Fennoscandian owls. Moreover, the ability to shift is dependent on the body size, hunting technique and availability of prey.

Of the nomadic species listed in Table 2, the Hawk Owl usually breeds in natural cavities, the Great Grey Owl on stick nests and the Snowy Owl on the ground. The hole-nesting habit of the Hawk Owl seems to disagree with Lundberg's (1979) hypothesis. However, all of these species are specialised on small rodents, so they have to search for an area with a good vole supply for wintering and breeding.

Male Tengmalm's Owls establish breeding territories and try to attract females (Korpimäki 1981). There are large inter-territorial differences in the breeding success (Korpimäki 1988b). As the reproductive success of males is dependent on the quality of territories that they occupy, a scarcity of high-quality territories with good nest-hole(s) selects for their residency (Lundberg 1979; Korpimäki 1981, 1986c, 1988b). In addition, the familiarity with the territory may make its defence, predator avoidance and foraging easier (Korpimäki, Lagerstrom & Saurola 1987). As females take part in the feeding of young only in the late nestling period (Korpimäki 1981),

detailed knowledge of the territory is much less advantageous for them than for males. Females are therefore free to shift their nest-sites according to the vole supply, but males cannot do so. Female biased seasonal movements were also documented for hole-nesting Hawk Owls in Norway (Byrkjedal & Langhelle 1986). Thus, the choice of nest-site seems to explain intraspecific differences in breeding dispersal.

In conclusion, I suggest that the hunting mode (quartering v. sit-and-wait technique), diet choice (specialised v. generalised) and nest-site (open nest v. hole-nest) are the three most important factors governing the seasonal migration patterns and dispersal tactics of Fennoscandian owls. This 'multifactor hypothesis' may explain both inter- and intra-specific differences described here better than do Lundberg's (1979) and Sonerud's (1986) hypotheses alone, though hunting technique and diet choice are partly intercorrelated.

Acknowledgements

I thank Hannu Pietiäinen for valuable comments on an earlier draft. My owl studies in South Ostrobothnia, western Finland, have been supported financially by the Finnish Cultural Foundation, the Oulu Student Foundation, the Jenny and Antti Wihuri Foundation, the Emil Aaltonen Foundation and the Academy of Finland.

References

Andersson, M. 1980. Nomadism and site tenacity as alternative reproductive tactics in birds. *Journal of Animal Ecology*, 49, 175–184.

Angelstam, P., Lindstrom, E. & Widen, P. 1984. Role of predation in short-term population fluctuations of some birds and mammals in Fennoscandia. *Oecologia*, 62: 199–208.

Angelstam, P., Lindstrom, E. & Widen, P. 1985. Synchronous short-term population fluctuations of some birds and mammals in Fennoscandia – occurrence and distribution. *Holarctic Ecology*, 8: 285–298.

Bairlein, F. 1985. Dismigration und Sterblichkeit in Suddeutschland beringter Schleierseulen (*Tyto alba*). *Vogelwarte*, *33*: 81–108.

Byrkjedal, I., & Langhelle, G. 1986. Sex and age biased mobility in Hawk Owls *Surnia ulula*. *Ornis Scandinavica*, 17: 306–308.

Carlsoon, B. G., Hornfeldt, B. & Loffren, O. 1987. Bigyny in Tengmalm's Owl *Aegolius funereus*: effect of mating strategy on breeding success. *Ornis Scandinavica*, 18: 237–243.

Cave, A.J. 1968. The breeding of the Kestrel, *Falco tinnunculus* L., in the reclaimed area Oostelijk Flevoland. *Netherlands Journal of Zoology*, 18: 313–407.

Dijkstra, C., Daan, S., Meijer, T., Cave, A.J. & Foppen, R.P.B. 1988. Daily and seasonal variations in body mass of the Kestrel in relation to food availability and reproduction. *Ardea*, 76: 127–140.

Erlinge, S. 1987. Predation and non-cyclicity in a microtine population in southern Sweden. *Oikos*, 50: 347–352.

Erlinge, S., Goransson, G., Hansson, L., Hodstedt, G., Liberg, O., Nilsson, I.N., Nilsson, T., van Schantz, T. & Sylven, M. 1983. Predation as a regulating factor in small rodent populations in southern Sweden. *Oikos*, 40: 36–52.

Erlinge, S., Goransson, G., Hogstedt, G., Liberg, O., Loman, J., Nilsson, I.N., Nilsson, T., von Schantz, T. & Sylven, M. 1982. Factors limiting numbers of vertebrate predators in a predator-prey community. *Transactions of the International Congress of Game Biologists, 14*: 261–268.

Exo, K.M. & Hennes, R. 1980. Beitrag zur Populations okologie des Steinkauzes *Athene noctua* – eine Analyse deutscher und niederlandischer Ringfunde. *Vogelwarte, 30*: 162–179.

Hagen, Y. 1965. The food, population fluctuations and ecology of the Long-eared Owl *Asio otus* L. in Norway. *Meddelelser fra Statens Viltundersokelser, 2*: 3–38.

Hansson, L. 1984. Predation as the factor causing extended low densities in microtine cycles. *Oikos, 43*: 255–256.

Hansson, L. & Henttonen, H. 1985. Gradients in density variations of small rodents: the importance of latitude and snow cover. *Oecologia, 67*: 394–402.

Hansson, L. & Henttonen, H. 1988. Rodent dynamics as community processes. *Trends in Ecology and Evolution, 3*: 195–200.

Henttonen, H. 1985. Predation causing extended low densities in microtine cycles: further evidence from shrew dynamics. *Oikos, 45*: 156–157.

Henttonen, H., Olsanen, T., Jortikka, A. & Haukisalmi, V. 1987. How much do weasels shape microtine cycles in northern Fennoscandian taiga? *Oikos, 50*: 353–365.

Holmgren, V. 1983. Skanska faglar: Invandring, forekomst och hackningsbiologi hos tornugglan. *Anser, 22*: 27–42.

Kalela, O. 1962. On the fluctuations in the numbers of arctic and boreal small rodents as a problem of production biology. *Annales Academiae Scientiarum Fennicae A IV, 66*: 1–38.

Korpimäki, E. 1981. On the ecology and biology of Tengmalm's Owl *Aegolius funereus* in Southern Ostrobothnia and Suomenselka, western Finland. *Acta Universitatis Ouluensis Series A. Scientiae Rerun Naturalium No. 118. Biologica 13*: 1–84.

Korpimäki, E. 1984. Population dynamics of birds of prey in relation to fluctuations in small mammal populations in western Finland. *Annales Zoologici Fennici, 21*: 287–293.

Korpimäki, E. 1985. Rapid tracking of microtine populations by their avian predators: possible evidence for stabilising predation. *Oikos, 45*: 281–284.

Korpimäki, E. 1986a. Seasonal changes in the food of the Tengmalm's Owl *Aegolius funereus* in western Finland. *Annales Zoologici Fennici, 23*: 339–344.

Korpimäki, E. 1986b. Diet variation, hunting habitat and reproductive output of the Kestrel *Falco tinnunculus* in the light of the optimal diet theory. *Ornis Fennica, 63*: 84–90.

Korpimäki, E. 1986c. Gradients in population fluctuations of Tengmalm's Owl *Aegolius funereus* in Europe. *Oecologia, 69*: 195–201.

Korpimäki, E. 1986d. Predation causing synchronous decline phases in microtine and shrew populations in western Finland. *Oikos, 46*: 124–127.

Korpimäki, E. 1986e. Niche relationships and life-history tactics of three sympatric *Strix* owl species in Finland. *Ornis Scandinavica, 17*: 126–132.

Korpimäki, E. 1987a. Dietary shifts, niche relationships and reproductive output of coexisting Kestrels and Long-eared Owls. *Oecologia, 74*: 277–285.

Korpimäki, E. 1987b. Selection for nest-hole shift and tactics of breeding dispersal in Tengmalm's Owl *Aegolius funereus*. *Journal of Animal Ecology, 56*: 185–196.

Korpimäki, E. 1987c. Timing of breeding of Tengmalm's Owl *Aegolius funereus* in relation to vole dynamics in western Finland. *Ibis, 129*: 58–68.

Korpimäki, E. 1987d. Clutch size, breeding success and brood size experiments in Tengmalm's Owl *Aegolius funereus*: a test of hypotheses. *Ornis Scandinavica, 18*: 277–284.

Korpimäki, E. 1987e. Sexual size dimorphism and life-history traits of Tengmalm's Owl: a review. *Biology and conservation of northern forest owls, symposium proceedings*, ed. by R.W. Nero, R.J. Clark, R.J .Knapton & R.H. *Hamre, 157–161. USDA Forest Service General Technical Report RM-142.*

Korpimäki, E. 1988a. Diet of breeding Tengmalm's Owls *Aegolius funereus*: long-term changes and year-to-year variation under cyclic food conditions. *Ornis Fennica, 65*: 21–30.

Korpimäki, E. 1988b. Effects of territory quality on occupancy, breeding performance and breeding dispersal in Tengmalm's Owl. *Journal of Animal Ecology, 57*: 97–108.

Korpimäki, E. 1988c. Factors promoting polygyny in European birds of prey – a hypothesis. *Oecologia, 77*: 278–285.

Korpimäki, E. 1988d. Effects of age on breeding performance of Tengmalm's Owl *Aegolius funereus* in western Finland. *Ornis Scandinavica, 19*: 21–26.

Korpimäki, E. 1989. Mating system and mate choice of Tengmalm's Owls *Aegolius funereus*. *Ibis, 131*: 31–50.

Korpimäki, E. & Lagerstrom, M. 1988. Survival and natal dispersal of fledglings of Tengmalm's Owl in relation to fluctuating food conditions and hatching date. *Journal of Animal Ecology, 57*: 433–441.

Korpimäki, E. & Norrdahl, K. 1989a. Avian predation on mustelids in Europe. 2: impact on small mustelid and microtine dynamics – a hypothesis. *Oikos, 55*: 273–276.

Korpimäki, E. & Norrdahl, K. 1989b. Predation of Tengmalm's Owls: numerical responses, functional responses and dampening impact on population fluctuations of microtines. *Oikos, 54*: 154–164.

Korpimäki, E. & Norrdahl, K. 1989c. Avian predation on mustelids in Europe 1: occurrence and effects on body size variation and life traits. *Oikos, 55*: 205–215.

Korpimäki, E. & Sulkava, S. 1987. Diet and breeding performance of Ural Owls *Strix uralensis* under fluctuating food conditions. *Ornis Fennica, 64*: 57–66.

Korpimäki, E., Lagerstrom, M. & Saurola, P. 1987. Field evidence for nomadism in Tengmalm's Owl *Aegolius funereus*. *Ornis Scandinavica, 18*: 1–4.

Linden, H. 1988. Latitudinal gradients in predator-prey interactions, cyclicity and synchronism in voles and small game populations in Finland. *Oikos, 52*: 341–349.

Lundberg, A. 1979. Residency, migration and a compromise: adaptations to nest-site scarcity and food specialisation in three Fennoscandian owl species. *Oecologia, 41*: 273–281.

Lofgren, O., Hornfeldt, B. & Carlsson, B.G. 1986. Site tenacity and nomadism in Tengmalm's Owl *Aegolius funereus* (L.) in relation to cyclic food production. *Oecologia, 69*: 321–326.

Mikkola, H. 1983. *Owls of Europe.* Calton, T & A D Poyser.

Nilsson, I.N. 1981. Seasonal changes in food of the Long-eared Owl in southern Sweden. *Ornis Scandinavica, 12*: 216–223.

Pershagen, H. 1969. Snow-cover in Sweden 1931–60. *Sveriges Meteorologiska och Hydrologiska Inst. Meddeleser Series A.* Nr. 5.

Pietiäinen, H. 1988. Breeding season quality, age, and the effect of experience on the reproductive success of the Ural Owl *Strix uralensis*. *Auk, 105*: 316–324.

Pietiäinen, H. 1989. Seasonal and individual variation in the production of offspring in the Ural Owl *Strix uralensis*. *Journal of Animal Ecology, 58*: 905–920.

Saurola, P. 1983. Movements of Short-eared Owl *Asio flammeus* and Long-eared Owl *A. otus* according to Finnish ring recoveries. *Lintumies, 18*: 67–71. [In Finnish with English summary.]

Saurola, P. 1987. Mate and nest-site fidelity in Ural and Tawny Owls. *In: Biology and conservation of northern forest owls, symposium proceedings*, ed. by R.W. Nero, R.J. Clark, R.J. Knapton & R.H. Hamre, 81–86. USDA Forest Service General Technical Report RM-142.

Saurola, P. 1989. Breeding strategy of the Ural Owl *Strix uralensis*. *In: Raptors in the modern world*, ed. by B.U. Meyburg & R.D.Chancellor, 235–240. Berlin, London & Paris, WWGBP.

Solantie, R. 1975. The areal distribution of winter precipitation and snow depth in March in Finland. *Ilmatieteen Laitoksen Tiedonantoja, 28*: 1–66.

Solantie, R. 1977. On the persistence of snow cover in Finland. *Ilmatieteen Laitoksen Tutkimuksia, 60*: 1–68.

Sonerud, G.A. 1986. Effect of snow cover on seasonal changes in diet, habitat, and regional distribution of raptors that prey on small mammals in boreal zones of Fennoscandia. *Holarctic Ecology, 9*: 33–47.

Sonerud, G.A. 1988. What causes extended lows in microtine cycles? *Oecologia, 76*: 37–42.

Sonerud, G.A., Solheim, R. & Prestrud, R. 1988. Dispersal of

Tengmalm's Owl *Aegolius funereus* in relation to prey availability and nesting success. *Ornis Scandinavica, 19*: 175–181.

Southern, H.N. 1970. The natural control of a population of Tawny Owls *Strix aluco. Journal of Zoology, 126*: 197–285.

Sulkava, S. 1966. Huuhkajan pesimaaikaisesta ravinnosta Suomessa. *Suomen Riista, 18*: 145–156.

Village, A. 1987. Numbers, territory-size and turnover of Short-eared Owls *Asio flammeus* in relation to vole abundance. *Ornis Scandinavica, 18*: 198–204.

Wallin, K. 1988. Life-history evolution and ecology in the Tawny Owl *Strix aluco*. Ph.D. thesis, Dept. of Zoology, University of Goteborg, Sweden.

Wallin, K., Jaras, T., Levin, M., Stranvik, B. & Wallin, M. 1983. Reduced adult survival and increased reproduction of Swedish Kestrels. *Oecologia, 60*: 302–305.

Wardhaugh, A.A. 1984. Wintering strategies of British owls. *Bird Study, 31*: 76–77.

Wijnandts, H. 1984. Ecological energetics of the Long-eared Owls (*Asio otus*). *Ardea, 72*: 1–92.

Willgohs, J.F. 1974. The eagle owl *Bubo bubo* (L.) in Norway. Part I. Food ecology. *Sterna, 13*: 129–177.

Asio owls and Kestrels in recently-planted and thicket plantations

A. Village

Village, A. 1992. Asio owls and Kestrels in recently-planted and thicket plantations. *In: The ecology and conservation of European owls*, ed. by C.A. Galbraith, I.R. Taylor and S. Percival, 11-15. Peterborough, Joint Nature Conservation Committee, (UK Nature Conservation, No. 5.)

The paper presents details of the population ecology of *Asio* owls and Kestrels *Falco tinnunculus*. Population densities were related to vole abundance in Scottish plantations. Some species are 'nomadic', enabling them to exploit the wide temporal and spatial fluctuations in prey. Others, such as the Kestrel, are residents and adopt a generalist feeding pattern enabling them to feed on alternative prey types when vole numbers are low.

Andrew Village, Natural Environment Research Council. Institute of Terrestrial Ecology, Monks Wood Experimental Station, Abbots Ripton, Huntingdon PE17 2LS.

Present Address: 4 Cedar Park, Stoke Bishop, Bristol BS91 1BW

Introduction

This paper summarizes previously published accounts of the population ecology of *Asio* owls and European Kestrels *Falco tinnunculus* in young conifer plantations in South Scotland between 1975 and 1979. New data from the same area collected during 1987 are also presented, in order to compare the breeding density and performance of owls and Kestrels before and after canopy closure. The information on owls was collected incidentally during the main study (of Kestrels), and is therefore confined to those aspects of data which could be most easily collected: breeding density and performance for Long-eared Owls *Asio otus* and number and territory size for Short-eared Owls *A. flammeus*. Throughout the paper, comparisons are drawn with Kestrels because all three species fed mainly on Short-tailed Voles, *Microtus agrestis* (Village 1981, 1982a, 1987).

Avian vole predators show a variety of responses to fluctuations in prey density (Village 1990). Many are able to turn to alternative prey items if voles are scarce. Their ability to do this depends on their hunting techniques. Those species that hunt mainly by sound have specialised morphological and behavioural adaptations which aid the efficient catching of small mammals. When small mammals are scarce, such species must either move away in search of a better food supply, suffer a reduction in breeding output or die. Predators such as Kestrels, which can feed on a variety of prey, should however be able to persist in poorer vole conditions, although they may also show enhanced densities and breeding productivity when voles are abundant.

The development of an even-aged conifer plantation is likely to result in marked changes in vole abundance. In the early stage after planting the rapid growth of grasses creates ideal Short-tailed Vole habitat, but this is gradually lost as the canopy closes. During the thicket stage, vole habitats are largely confined to gaps between compartments, areas of poor tree growth or unplanted areas alongside roads and streams. Unpredictable combinations of changes in the distribution of vole habitats and annual fluctuations in the vole cycle occur. For example young plantations may sometimes have few voles, whilst the reverse might be true for patches of thicket plantations. The response of predators to vole distributions within particular forest developments is not, therefore, easy to predict and has received rather little attention despite the major habitat changes brought about by commercial forestry over the last twenty years.

The study area and methods are described in detail elsewhere (Village 1981 (Long-eared Owls); 1982a , 1982b, 1986, 1990 (Kestrels); 1987 (Short-eared Owls)) and are merely summarised here. The study area was in Eskdalemuir Forest, south Scotland, and covered about 100 km². Plantations comprised 60% of the forest area in 1975, the majority of which was less than eight years old. The land outside the plantations was mainly sheepwalk, grazed by sheep and some cattle. Mature trees were mainly conifers, occurring as single trees or small shelterbelts, mostly in the valleys. Disused Carrion Crow *Corvus corone* nests in these trees provided most of the natural nest sites for Long-eared Owls and Kestrels.

Breeding density and performance were recorded for Kestrels and Long-eared Owls by searching all the woods for signs such as droppings, pellets, fresh kills or evidence of old crow nests. A site was assumed to be occupied if a pair of birds was seen or a scraped nest found; non-laying was assumed if there was no evidence that scraped nests had contained eggs. Nests were visited as often as necessary to record laying date, clutch size and number of young fledged. In 1976 and 1987 artificial stick-nests were erected in some woods to increase the number of potential sites in the area. These nest sites were readily used by owls and Kestrels. These artificial nest sites ensured that the density of potential nest sites was roughly constant from year to year.

An index of the population densities of Kestrels and Short-eared Owls was calculated from roadside counts. All birds seen during drives along forest tracks and minor roads were recorded and the number seen per km driven in sheepwalk. The number of birds observed in new plantings (0–5 year) and established planting (6–10 year) was also recorded by totalling the figures for autumn (October-November), winter (December-February) and spring (March-May).

Vole numbers were assessed by snap-trapping at 20 sites (17 in young plantation and 3 in sheepwalk) in April and October from 1975 to 1979 (Village 1981, 1982a). The number per trap site was used as an index of abundance in autumn and spring, winter values being estimated by linear interpolation between these two values (Village 1982b).

The author visited the area several times between March and June 1987 in order to estimate the breeding density of Long-eared Owls and Kestrels. Most of the plantations had a closed canopy, although the trees held few crow nests. It was therefore possible to find most of the likely nest sites and be reasonably sure of finding all the breeding pairs. About 7 km² of sheepwalk in the middle of the forest had recently been planted, and this may have made parts of the adjoining forest more suitable for vole predators than would normally have been the case. No roadside counts of Short-eared Owls were made in 1987, and information was confined to the other two species. Vole numbers continued to be monitored at Eskdalemuir as part of a wider survey in the region, and the estimate used for 1987 was that given for young plantations in the Eskdale region (Taylor *et al.* 1988).

Results

Short-eared Owls in young plantations 1975–78

Short-eared Owls were always much more abundant in the plantation than in sheepwalk and were only

seen using the latter habitat when vole numbers were increasing in 1978. This paralleled the greater abundance of voles in the planted areas. Fluctuations of owl and vole numbers over time within young plantations were significantly correlated (Figure 1). The same was true for Kestrels, although they were more frequent in sheepwalk than were Short-eared Owls (Village 1987).

The seasonal changes in owl and vole numbers were not closely correlated, owls being more abundant relative to voles in spring than in autumn or winter (Fig. 1b). The same was true for Kestrels (Village 1982b), and in both species this seemed to be because territories were occupied by pairs in spring but by single birds in winter (Village 1982b, 1987). Vole numbers were at their annual low in spring, and owl territories and Kestrel home ranges were about the same size as would be expected for single birds at low vole densities in autumn.

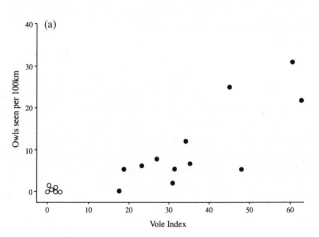

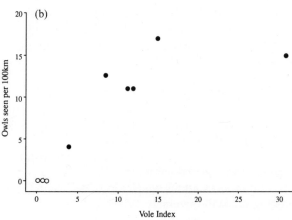

Figure 1. Relationship of Short-eared Owl numbers to vole numbers in (a) autumn and winter and (b) spring, at Eskdalemuir 1975–78. Open circles = Sheepwalk, closed circles = young plantation. Correlations were significant for young plantation in (a) (r = 0.87, P < 0.01) and for both habitats combined in (b) (r = 0.82, P < 0.01). Re-drawn from Village (1987).

12

Twenty one breeding Short-eared Owls were trapped and wing-tagged during 1976. However, only one of these 21 birds was present the following year. Tagging had occured just prior to a decline in vole numbers. This can be compared to three out of seven owls, marked during 1977, subsequently being recorded after a year when vole numbers were increasing. After the initial vole decline, owls moved long distances and were recovered up to 1700 km S and 500 km NNW during the following year. One tagged owl was recorded breeding in Orkney (420 km NNW) in 1977.

Long-eared Owls and Kestrels in young plantations 1976–79

Long-eared Owls bred at higher densities, laid earlier and produced larger clutches in spring when voles were abundant than when they were scarce (Village 1981). The variables more closely associated with vole abundance were those determined early in the season i.e. the nearest-neighbour distance between pairs, the proportion of pairs that produced eggs, mean laying date and mean clutch size. The proportion of clutches that were deserted and the proportion of broods surviving to fledging age were not correlated with spring vole numbers. A higher proportion of hatched young survived to fledge in poor vole years than in good ones. This latter trend was against the expected direction and may have been spurious. Alternatively, only 'better' owls, those most likely to rear young to independence, might have hatched young in poor vole years. Similar relationships between spring vole numbers and breeding density, laying date and clutch size were apparent in Kestrels, which were also more likely to fail in the early stages of breeding than with young (Village 1986).

Long-eared Owls and Kestrels in thicket plantation, 1987

Vole numbers were at a peak in 1987 (Taylor *et al.* 1988) and the vole index was higher than in any previous spring from 1976 to 1979. Both Long-eared Owls and Kestrels bred early, laid large clutches and had large broods (Table 1). The most reliable data on breeding performance obtained were mean laying dates for Kestrels and mean clutch sizes for Long-eared Owls. Both these were at levels that would have been predicted from the same vole abundance in young plantation (Figs 2 and 3).

For compatability with earlier papers, breeding density in Kestrels was expressed as pairs per 100 km² and in Long-eared Owls as nearest-neighbour distance of pairs (which is inversely related

Table 1. Breeding density and performance of Long-eared Owls and Kestrels at Eskdalemuir, 1976–79 and 1987.

a. Long-eared Owls

Year	Spring Vole index	Mean nearest-neighbour distance (km)	Mean clutch
76	13.6	1.37	3.2
77	11.5	1.03	4.0
77	4.0	1.57	2.5
78	30.5	0.43	4.7
78	1.6	1.45	4.8
79	29.0	0.94	5.5
79	18.2	1.41	4.1
87	26.0	1.46	5.3

b. Kestrels

Year	Spring Vole Index	Density of pairs (per 100 km²)	Mean laying date (1 = April 1)
76	13.6	30	28
77	5.8	27	41
78	16.1	35	28
79	20.0	36	32
87	26.0	21	25

Data for 1976–79 from Village (1981, 1990). Long-eared Owl results were divided into low and high areas in some years.

KESTRELS

LAYING DATE

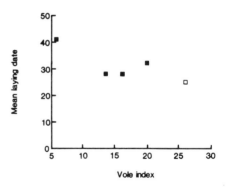

Figure 2. Mean laying date of Kestrels at Eskdalemuir in relation to spring vole abundance for 1976–79 (solid squares) and 1987 (open square).

to density). In both species, breeding density was lower than would have been expected in young plantation at the same vole density (Figs 4 and 5).

Discussion

In all three raptors, population densities in young plantations were related to vole abundance. Vole numbers were low in spring 1977, and there were few Short-eared Owls in the area. The recoveries and sightings of adults wing-tagged in the previous breeding season suggested that movements out of the

LONG-EARED OWLS
MEAN CLUTCH-SIZE

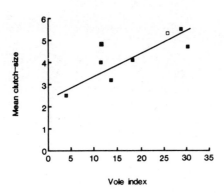

Figure 3. Mean clutch size of Long-eared Owls at Eskdalemuir in relation to spring vole abundance in 1976–79 (solid squares) and 1987 (open square). The line is the fitted linear regression for the 1976–79 data, from Village (1981).

KESTRELS
BREEDING DENSITY

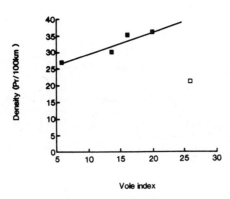

Figure 4. Breeding density of Kestrels at Eskdalemuir in relation to spring vole abundance in 1976–79 (solid squares) and 1987 (open square). The line is in the linear regression fitted to the solid squares.

area were a major cause of the decline, and some individuals dispersed long distances to breed elsewhere. Such 'nomadic' behaviour occurs in other northern temperate species that are heavily dependent on voles (Village 1990), and may enable them to effectively exploit the wide temporal and spatial fluctuations in their prey.

Kestrels were also scarce in spring 1977. However, no breeding adults were subsequently reported breeding long distances away. There was therefore no evidence to suggest truly nomadic behaviour. Kestrels in the area were partially migrant (Village 1982b). When vole numbers were low throughout an area, adult Kestrel mortality during the winter was high. New immigrants passing through the area during the spring may also have been unable to settle as a result

LONG-EARED OWLS
NEAREST-NEIGHBOUR DISTANCE

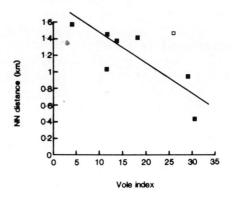

Figure 5. Mean nearest-neighbour distance of Long-eared Owls in relation to spring vole abundance in 1976–79 (solid squares) and 1987 (open square). The line is the linear regression fitted to all solid squares (after Village 1981).

of the poor food supply. About 17% of 207 wing-tagged Kestrels were known to have migrated back to an area after a previous breeding attempt, and 11% remained for a least one whole year in an area (Village 1985). Kestrels were thus more tied to an area than Short-eared Owls, and were able to persist, albeit in reduced numbers, in poor vole years. They were also able to exploit sheepwalk, unlike the owls, partly because they took a wider variety of prey (Village 1982a).

The population regulation of Long-eared Owl numbers was harder to define as few breeding adults were tagged. The density of Long-eared Owl pairs was correlated with vole abundance and this correlation was reflected as a change in the total owl population size. Some single birds may have been present, but characteristically breeding or roosting sites showed no visible signs of occupation by owls when voles were scarce. Some adults may have moved elsewhere when voles declined. For example, one breeding female, tagged in 1977, was found shot on a nest some 48 km away, two years later. Decline in breeding owl density may be due in part to pairs being unsuccessful at laying, as a result of having persisted in an area where vole numbers were too low to permit breeding. This suggests less nomadic behaviour in this species than in Short-eared Owls, though non-breeding by the latter species could not be excluded. Clearly, there is scope for a more detailed study of Long-eared Owls involving marked adults.

Density and breeding performance in thicket plantations

The early laying and large clutches of Kestrels and Long-eared Owls at Eskdalemuir in 1987 suggest that these species are able to successfully exploit the available vole habitat in thicket plantation. More data are needed from such habitats in poor vole years, but the available evidence suggests that breeding performance would vary with vole density with a similar pattern to populations in recently planted forests.

The main difference between young and thicket plantations was the lower breeding density of owls and Kestrels in the more mature forest compared with young plantations. There are two possible reasons for this. Firstly, the greatly reduced area of habitat suitable for supporting voles may have reduced the carrying capacity of the forest by about 30–40% as compared with that of the early post-planting stage. Secondly, competition from Tawny Owls *Strix aluco*, which may be better adapted to thicket plantation than either Kestrels or Long-eared Owls, may have led to reduced breeding densities.

Tawny Owls were unusual at Eskdalemuir in the 1970s, the number of pairs found ranging from one in 1977 to six in 1979. This compares with the 13 pairs found in 1987. The mean nearest-neighbour distance of all Long-eared and Tawny Owls together in 1987 was 1.12 km, which was closer to the distance predicted for Long-eared Owls alone at the vole density in young plantation (Figure 5). Tawny Owls are larger than the other species which may have excluded them from using certain nest sites. Several Long-eared Owls and Kestrels nested in artificial baskets in 1987, but more detailed work is needed to determine whether the low densities of these species in thicket plantation are due mainly to a lack of vole habitat, direct competition for feeding sites with Tawny Owls, or competition for scarce nesting sites.

Acknowledgements

I thank the Economic Forestry Group for giving me access to Eskdalemuir Forest, and Ronnie Rose for his help and support in many ways. Eskdalemuir Observatory kindly provided accommodation in 1987. Nigel Charles, Iain Taylor and Ian Langford generously allowed me to use their vole data.

References

Taylor, I.R., Dowell, A., Iriving, T., Langford, I.K. & Shaw, G. 1988. The distribution and abundance of the Barn Owl, *Tyto alba*, in south-west Scotland. *Scottish Birds 15*: 40–43.

Village, A. 1981. The diet and breeding of Long-eared Owls in relation to vole numbers. *Bird Study, 28*: 215–224.

Village, A. 1982a. The diet of Kestrels in relation to vole abundance. Bird Study, *29*: 129–138.

Village, A. 1982b. The home range and density of Kestrels in relation to vole abundance. *Journal of Animal Ecology, 51*: 413–428.

Village, A. 1985. Turnover, age and sex ratios of Kestrels Falco tinnunculus in south Scotland. *Journal of Zoology, London (A), 206*: 175–189.

Village, A. 1986. Breeding performance of Kestrels at Eskdalemuir, south Scotland. *Journal of Zoology, London (A) 208*: 367–378.

Village, A. 1987. Numbers, territory-size and turnover of Short-eared Owls *Asio flammeus* in relation to vole abundance. *Ornis Scandinavica, 18*: 198–204.

Village, A. 1990. *The Kestrel*. Calton, Poyser.

The population ecology and conservation of Barn Owls *Tyto alba* in coniferous plantations

I.R. Taylor, A. Dowell and G. Shaw

Taylor, I.R., Dowell, A. & Shaw, G. 1992. The population ecology and conservation of Barn Owls *Tyto alba* in coniferous plantations. *In: The ecology and conservation of European owls*, ed. by C.A. Galbraith, I.R.Taylor and S. Percival, 16-21. Peterborough, Joint Nature Conservation Committee. (UK Nature Conservation, No. 5.)

Barn Owls breeding in coniferous plantations in southern Scotland preyed mostly on Field Voles *Microtus agrestis* which underwent cyclic changes in abundance. Laying dates, clutch size and production of young were significantly related to vole abundance as was the number of pairs of owls attempting to breed. The number of owls breeding in plantations was increased significantly by the provision of artificial nest sites.

I.R. Taylor, Institute of Cell, Animal and Population Biology, University of Edinburgh, Zoology Building, West Mains Road, Edinburgh EH9 3JT

A. Dowell and G. Shaw, Forestry Commission, Forest District Office, Creebridge, Newton Stewart, Wigtownshire DG8 6AJ

Introduction

Coniferous plantations, especially in their early stages of growth, provide important habitat for many diurnal and nocturnal birds of prey, several of which (e.g. Hen Harrier *Circus cyaneus*, Long-eared Owl *Asio otus*, Short-eared Owl *Asio flammeus* and Barn Owl *Tyto alba*) are uncommon in Britain. Plantations now contribute significantly to their conservation, partly through protection from persecution which, although illegal is still widespread, and partly through the provision of prey rich hunting areas (Petty, this symposium; Village 1981, 1982, 1983, 1987, this symposium). The increase in biomass of vegetation following fencing, drainage and planting is associated with substantial increases in the density of Field Voles *Microtus agrestis*, a major prey species for many of the birds of prey and several mammalian predators. Vole densities are highest in plantations on mineral soils where *Agrostis* spp. are dominant and lowest on blanket peat where *Calluna vulgaris* is the dominant species. They remain accessible to most predatory birds until canopy closure, 10–15 years after planting. In closed forest, vole densities remain higher in the rides than on unplanted ground and clear felled and restocked sites also have high densities (Charles 1981; Petty 1988). Thus, plantations have the potential to offer suitable long-term habitat for many predatory species.

This paper considers the population ecology and conservation of Barn Owls in coniferous plantations. It examines (1) the role of food supply in determining the productivity of the owls, (2) the significance of both food supply and nest sites in determining the numbers breeding in plantations and (3) the relative abundances of field voles along forest edges, at the boundary with sheepwalk, an aspect of habitat not previously investigated. The nature of this boundary varies but most often a narrow strip is left between exclosure fences and trees which in the absence of grazing develops rank grassland or heathland vegetation. As this zone persists throughout the planting cycle it is potentially an important hunting area for predators. Studies in farmland have shown such woodland edges to be the most important hunting habitat for Barn Owls (Taylor in prep.)

Study areas

The study involved two areas in south Scotland: the Esk valley in east Dumfries and the Cree valley north and west of Newton Stewart, Galloway. In both areas, former rough grazing and hill land has been extensively afforested mostly with Sitka *Picea sitchensis* and Norway Spruces *P. abies*. The Cree area has a more mature and varied forest with a mosaic of recently afforested and restocked sites and all stages of forest development represented. The Esk area forests are mostly younger (less than 20 years) with only limited areas of second generation plantation.

Methods

The study began in 1979 in both study areas. At the start, the areas were searched intensively for suitable Barn Owl nest sites in buildings, trees and cliffs. Thereafter, the nest sites were examined each year up to 1988 to determine the number available to the owls, the number used for breeding, laying dates, clutch size, hatching success and the number of young fledged. The breeding season extended from March to September and normally four to six visits were needed for each nest site over this period.

In the Esk area, samples of pellets were collected from each nest throughout the breeding season and analysed to provide details of diet. Normally, 30–50 pellets were needed to give a sample size of about 150 prey items for each site. The relative abundances of Field Voles and Common Shrews *Sorex araneus* were also estimated each year in the spring (late March/early April) by trapping, using methods developed by Charles (1981). This involved a grid of 24 trapping stations selected randomly with two treadle operated snap traps at each, set in sample sites of two to four year old Sitka Spruce plantations. Each year, six such sites spread over the study area were sampled. In each, trapping was undertaken continuously over five nights with traps examined and reset every day. The total numbers of voles and shrews caught were totalled and the mean of such totals for the six sites were used as the annual index of abundance.

The abundance of Field Voles in the forest edge habitat was examined in the Esk area in 1981 comparing edge habitat, adjacent sheepwalk and closed forest. Six sample sites were selected with a 4–6 m wide rank grassland strip alongside 15–25 year old Sitka Spruce plantations and fenced off from open sheepwalk. A line of 20 trapping stations, 10 metres apart with two snap traps at each was set up in each of the sheepwalk, grass edge strip and closed plantation. The traps were operated for five 24-hour periods.

From 1985 to 1988, in the Cree study area, we tested the hypothesis that the availability of nest sites limited the number of Barn Owls breeding in coniferous plantations by artificially increasing the number of sites available. Plastic drums of 18–20 gallon capacity were made suitable as nest sites by cutting a 10×10 cm entrance hole near the top and providing a layer of wood shavings as a nesting base. These were attached at a height of 4–5 m to trees throughout the area. By 1988, drums were in place at 57 sites; 13 adjacent to plantations less than six years old, 19 along sides and firebreaks in closed forest, 20 along forest edges adjacent to farmland and five at clearfell sites. Two drums were placed at each site such that if one were used by Tawny Owls *Strix aluco* the other was still available to Barn Owls. The drums were checked for occupancy three times each year, between April and July. A drum was counted as occupied for breeding if at least one egg had been laid.

Results

Effects of food supply

The diet of Barn Owls in the breeding season (April to July) was determined by pellet analysis for a total of 52 nests in the Esk study area during eight of the ten study years. This represented 80% of all nesting attempts. At all nests, the diet consisted almost entirely of two prey species, the Field Vole and the Common Shrew, but there were large annual variations in the relative proportions of these species. The percentage of voles taken each year was significantly correlated with the abundance of voles in the plantations (Figure 1). When voles were at their most abundant, they constituted on average just over 75 percent of the items in the owls' diet, declining to just under 30 percent when they were least abundant. The opposite trend occurred with common shrews; their contribution to the diet increased significantly as vole populations declined ($r = 0.91$, $p < 0.01$).

The importance of Field Voles is better assessed by examining their contribution to the diet by weight rather than number. Voles captured by the owls weighed on average just over three times as much as Common Shrews. Thus, even when voles were least numerous in the diet, they still contributed 58 percent by weight, on average, and at their most numerous, 91 percent.

Other prey appeared much less frequently in the pellets. Common Frogs *Rana temporaria* were taken occasionally in early spring and in summer a small number of Pipstrelle Bats *Pipistrellus pipistrellus*, Pigymy Shrews *Sorex araneus* and newly fledged song-birds (mostly Meadow Pipit *Anthus pratensis*) were recorded. These items together never

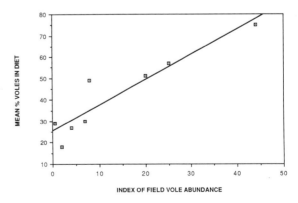

Figure 1. The percentage of Field Voles *Microtus agrestis* in the diet of Barn Owls during the breeding season (April to August) in relation to yearly changes in the abundance of voles in plantations, Esk study area ($r = 0.94$, $p < 0.01$).

contributed more than 2 percent by weight to the total diet.

The annual sampling of Field Voles in the Esk area showed that their numbers varied greatly from year to year approximating to a cyclical pattern with peak abundances in 1981, 1984 and 1987 and lowest abundances in 1983, 1985 and 1988. In 1981, the year of greatest numbers, voles were 88 times more abundant than in 1985, the year of lowest numbers. Vole populations were not monitored quantitatively in the Cree study area but large scale changes were evident from the number of runs and other signs. Changes in this area followed a similar pattern to those in the Esk area with the exception of 1988, when numbers remained high in the Cree area but 'crashed' in the Esk area. Common Shrew numbers in the Esk area varied in synchrony with vole numbers ($r = 0.70$, $p < 0.05$) but the amplitude of the fluctuations in shrew numbers was not as great as that of the voles, with only a four-fold difference in the number trapped between years of highest and lowest abundance.

The owls thus experienced large variations in the abundance of their main prey species. This had a marked effect on all aspects of their breeding performance. The mean dates of the start of laying in Esk were significantly related to vole abundance ($r = 0.76$, $p < 0.05$) with a difference of almost 30 days between years of highest and lowest abundance. Mean clutch size was also significantly related to vole abundance ($r = 0.80$, $p < 0.02$) as was the mean number of young successfully fledged (Figure 2) which ranged from just over two per pair at the lowest vole abundances to almost six per pair at the highest vole abundances.

A detailed analysis of mortality patterns and rates will be presented elsewhere (Taylor in prep.). A brief summary of the main conclusions is given here as these are highly pertinent to conservation planning. Adult mortality occurred mostly from December to February. Starvation was a major cause of death and mortality rates were significantly higher (51.4%) under conditions of low vole abundance than under conditions of high vole abundance (27.6%, $p < 0.01$). Mortality rates also increased significantly with increasing altitude ($r = 0.97$, $p < 0.01$).

The number of pairs of Barn Owls that attempted to breed in the Esk area plantations varied from a minimum of two to a maximum of fourteen. This annual variation was significantly related to yearly changes in vole abundance (Figure 3).

Effect of nest site availability

Within the study forests Barn Owls were totally dependent on buildings for nest sites. These were mainly shepherds' or farm workers' cottages, abandoned before or during the conversion to forestry. In most cases, the birds chose to nest in the attic space or between the floorboards and ceiling, but occasionally they used chimney stacks. No pairs were found breeding in tree hole sites. This did not represent a preference for buildings but rather a lack of suitable nest holes in trees. Large deciduous trees were uncommon within the study forests and none were found with cavities large enough for Barn Owls. Disused buildings were also occupied by unmated birds in summer and for roosting outside the breeding season.

In both study areas, when the owl populations were at their highest, the number of breeding pairs came close to the number of nest buildings available. In the Esk area in 1981, there was only one surplus site. In the Cree area in 1988 only one building was unused (Table 1). This suggests that the number of pairs nesting within the plantations might be limited by available sites and raises the possibility of increasing

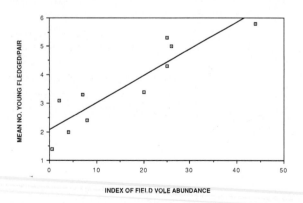

Figure 2. Mean number of young fledged by Barn Owls nesting in plantations in relation to yearly change in the abundance of Field Voles *Microtus agrestis*. Esk study area ($r = 0.91$, $p < 0.01$).

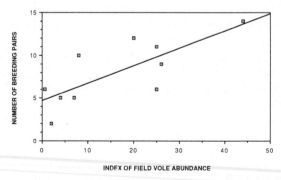

Figure 3. Yearly changes in the number of pairs of Barn Owls breeding in plantations in relation to changes in Field Vole abundance. Esk study area ($r = 0.77$, $p < 0.01$).

Table 1. Number of 'natural' nest sites (buildings) available and used by Barn Owls for breeding in the Esk and Cree study areas and the number of nest boxes available and used in the Cree area.

	79	80	81	82	83	84	85	86	87	88
Esk Study Area										
Natural sites available	17	16	15	16	17	17	16	14	12	10
Natural sites used	12	11	14	10	5	9	6	2	6	5
Cree Study Area										
Natural sites available	8	8	8	7	7	6	7	7	7	7
Natures sites used	6	5	5	4	5	4	5	5	5	6
Nest box sites available	–	–	–	–	–	–	26	32	49	57
Nest box sites used	–	–	–	–	–	–	3	4	9	29

Table 2. Number of artificial nest sites available and used by Barn Owls for breeding separated according to habitat type in the Cree Area forests in 1988.

	Available	Used for Breeding	%
Afforested sites (<6 years old)	16	10	76.9
Restocked sites	5	2	40.0
Rides and firebreaks in closed forest	19	8	42.2
Forest/farmland edge	20	9	45.0

their number by providing nest boxes. Also, as it would not be possible to justify the considerable expenditure that would be needed to maintain the buildings currently in use, existing nest sites will eventually disappear. In the Esk forests, 41% of buildings available to the owls in 1979 had become unsuitable by 1988. Seven of ten remaining buildings were also deteriorating rapidly and could not be expected to survive more than five years. A similar pattern was apparent throughout south Scotland. Thus, if Barn Owls are to be maintained in these plantations even at their present levels, artificial nest sites will have to be provided.

In the Cree area from 1985 to 1988 the number of available nest sites was increased to test whether Barn Owls would accept artificial sites in the forest and whether the number of buildings had previously limited the number of pairs nesting. Between 1979 and 1988, the number of pairs nesting in buildings was consistent at four to six. In 1985, 26 artificial sites were available and three (11.5%) were used for breeding. This increased to four out of 32 (12.5%) in 1986, nine out of 49 (18.4%) in 1987 and 29 out of 57 (50.9%) in 1988 (Table 1). The pattern of build up was probably related to vole abundance; 1985 and 1986 were years of low vole abundance whereas 1987 and 1988 had high vole populations. The much greater increase in 1988 probably arose from a high breeding performance in 1987, followed by a high overwinter survival and recruitment to the breeding population. Artificial nest sites were available alongside young plantations, restocked sites, rides and firebreaks in closed forest and at the forest/farmland edge. The sample size in each category was determined mainly by the availability of each habitat type within the study forests. Barn Owls attempted to breed at all four locations and although they made slightly more use of sites alongside young plantations, there was no significant difference between the use made of this habitat and any of the others (Table 2). Pairs nesting in the drums in 1988 had a mean clutch size of 4.9 (range 3–7) and fledged

on average 2.1 (range 0–6) young per pair. This is less than might be expected in a high vole year (Figure 2). However, of 24 breeding birds whose age was known from ringing, 19 (79%) were in their first year. In the Esk area, first year breeders produced significantly fewer young on average than older birds (Taylor in prep.). Such differences have also been identified in other predatory birds (Newton 1979). In the Esk area, first year birds never accounted for more than 40% of the breeding population. The relatively low average output in the artificial sites may thus have arisen partly from the highly skewed age distribution of the birds using them.

In 1988, eight of the artificial sites were occupied by Tawny Owls but there were no obvious signs of adverse interactions between this species and Barn Owls. Over the four year period that artificial sites were available, there was no case of Tawny Owls taking over a site previously occupied by Barn Owls. In 1988, there were 22 sites unoccupied by either species so it seems unlikely that Tawny Owls limited the number of Barn Owls nesting.

Abundance of Field Voles in forest edge habitat

A comparison of the relative abundance of Field Voles in open sheepwalk, forest edge and closed forest was made in 1981, a year of peak vole populations. Attempts were made to repeat the comparisons in years of low vole abundance but so few were trapped that analysis was not possible. In 1981, a mean of 30.5 Field Voles were trapped per sample site along the enclosed forest edges compared with 0.7 within the closed forest and 2.7 on the adjacent sheepwalk. Forest edges should thus be relatively rich hunting areas for Barn Owls (Table 3).

Discussion

This study has shown that coniferous plantations can provide suitable habitat for Barn Owls and that the number of breeding pairs can be increased substantially by the provision of artificial nest sites.

Table 3. Comparison of mean number ($\pm$SE) of Field Voles captured along grassy forest edges (4–6 m wide) with adjacent sheepwalk and closed forest plantation in 1981. Means from six sample sites in Esk study area. Grass edge vs sheepwalk, $p < 0.05$, Wilcoxon test.

	Closed forest	Grass edge	Sheepwalk
Mean no. Field Voles trapped per sample site	0.7 ± 0.3	30.5 ± 2.5	2.7 ± 0.8

However, as the study was carried out in only one area of Britain, it is important to consider the specific characteristics of that area and the extent to which the conclusions are applicable to other regions.

All the study forests were established on land that was previously predominantly grassland; either marginal or hill land managed for sheep production. Thus during the establishment phase and along forest rides, firebreaks and edges, the main vegetation form was rank grassland. The main prey species for Barn Owls, the Field Vole, has a preferred diet of grass leaves and stems, particularly species with high nutrient content and digestibility such as *Agrostis* spp., *Festuca* spp., *Anthoxanthum odoratum* and *Dactylis glomerata* (Summerhayes 1941; Hansson 1971a, 1971b; Evans 1973; Ferns 1976). These are some of the main grasses that are encouraged by management for sheep production (Eadie 1970; Gething et al. 1973). Hence, the long grassland that develops following exclosure and planting of former sheepwalk, by providing both food and cover, is often good quality habitat for Field Voles and high densities can be attained. Plantations established over more acidic peaty soils with predominantly heathland, moorland or bog vegetation, where the preferred vole food species are less abundant, support much lower densities (Hansson 1971b; Charles 1956, 1981). Thus the provision of Barn Owl nest boxes is most likely to be successful where plantations have been established on former grassland or where rich grassland has developed as a result of drainage following afforestation.

Altitude is another important consideration, as Barn Owl annual mortality rates were greater at higher altitude. Most losses occurred in winter and the altitude effect probably arose from a combination of increased food requirements resulting from lower temperatures and greater exposure, and deteriorating conditions for hunting resulting from increased rainfall, snowfall, depth and duration of snow cover and windspeed. All of these aspects of weather are altitude dependent (Chandler & Gregory 1979). Consequently, sites higher than about 200 metres above sea level in south Scotland were occupied much less consistently than lower sites. The Cree study area, where the nest box trial was successful, was mostly within an altitudinal range of 50 to 200 metres above sea level. The optimum altitude range

will probably vary across the country according to local climate, but in the absence of a more detailed understanding of the mechanisms underlying the effects of altitude on Barn Owls, it is not possible at present to be more precise.

A third important factor likely to influence the success of nest box provision in plantations, at least initially, is the density of Barn Owls breeding in the surrounding area and the distance to the nearest high density populations. In both south Scotland study areas, about 95% of ringed birds dispersed less than 15 km between their natal sites and their breeding sites (Taylor unpublished, Shaw & Dowell unpublished). South-west Scotland retains a relatively high density of Barn Owls (Taylor et al. 1988). The nest boxes erected in the Cree area plantations were colonised mainly by young produced by birds nesting close by on farmland. Even so, the occupation of the boxes did not occur immediately, but was delayed for three to four years, coinciding with the peak of the vole cycle and subsequent increase in the number of owls surviving and available for recruitment to the local population.

Assuming the above criteria have been met, forest structure is the main remaining factor to be considered. As Barn Owls need open areas supporting populations of Field Vole, large expanses of closed forest are of little or no value. The Cree area forests offered a mosaic of forest edge, newly afforested areas, restock sites, firebreaks and rides and closed forest. It is planned that most commercial forests in Britain will eventually move away from the large even aged plantations characteristic of the 1960s and 1970s towards the structure shown in the Cree forests (Hibbard 1985; Ratcliffe & Petty 1986). Barn Owls utilised nest boxes sited alongside forest edges, restock sites, newly afforested sites and rides. Restock sites in south Scotland and northern England tend to develop grassland supporting good Field Vole populations (Petty 1987) but this is probably dependent upon a high level of browsing from deer which prevents the establishment of tree seedlings and forbs (Ratcliffe & Petty 1986). In other areas such as north Wales, in the absence of deer, a diverse scrub layer tends to develop which would be less suitable for barn owls (Currie & Bamford 1981; Bibby et al. 1986). Similarly, the grass strips that develop along forest edges, shown here to hold dense

Field Vole populations, are probably also dependent on deer browsing and might not be maintained without management, such as periodic cutting, in areas with low deer populations.

Further research is needed to find out in greater detail how Barn Owls use such habitats as forest edges, firebreaks and restock sites. It would be useful to know, for example, if there is an optimum edge width or size of restock site and how much of each habitat type is needed for each pair of owls. More precise habitat management recommendations or advice on the siting of nest boxes cannot be made until such information becomes available.

In this study we used plastic drums as nest boxes, mainly because they were lightweight, waterproof and inexpensive which meant that large numbers could be erected for the trial. In 1988, a high vole year, Barn Owls nesting in these drums fledged 2.1 young per pair, somewhat lower than might be expected. It was suggested that this might have arisen from a very high percentage (79%) of first year breeders in the population, but we cannot exclude the possibility that the size of the boxes may have been a contributory factor (Korpimäki 1985). Further trials to compare the birds' productivity in relation to nest size would be useful. However, the use of plastic drums remains a convenient initial means of attracting the owls to breed in plantations, even if it becomes desirable to replace them with more substantial nest sites at a later stage.

Acknowledgements

It is a pleasure to thank Mr D.J. Perry, former Assistant Conservator for South Scotland and Mr S. Petty, Wildlife and Conservation Research Branch, for their enthusiastic support and advice for the work in the Cree study area. Many people helped with vole trapping but, in particular, we thank Dr A. Chaudhry, P. Bell and I.K. Langford. Thanks also to Drs M. Marquiss, I. Newton, A. Village, W.N. Charles and to J.Y. Ogilvie, J. Hamilton, M. Osborne, S. Abbott, F. Slack, R. Rose, Buccleuch Estates, Economic Forestry Group and Tillhill Forestry. A very special thanks to Tom Irving who helped in many ways and to Anne Aitken for typing the manuscript. The work was funded by the Nature Conservancy Council, World Wide Fund for Nature, Natural Environment Research Council, Forestry Commission, Edinburgh University and the senior author.

References

Bibby, C.J., Phillips, B.N. & Seddon, A.J.E. 1985. Birds of restocked conifer plantations in Wales. *Journal of Applied Ecology, 22*: 619–633.

Chandler, T.J. & Gregory, S. 1979. *The Climate of the British Isles.* Longman, London.

Charles, W.N. 1956. The effect of a vole plague in the Carron Valley, Stirlingshire. *Scottish Forestry, 10*: 201–204.

Charles, W.N. 1981. Abundance of field vole *Microtus agrestis* in conifer plantations. *In: Forest and Woodland Ecology,* ed. by F.T. Last & A.S. Gardiner, 135–137. Institute of Terrestrial Ecology, Cambridge.

Currie, F.A. & Bamford, R. 1981. Bird populations of sample pre-thicket forest plantation. *Quarterly Journal of Forestry, 75*: 75–82.

Eadie, J. 1970. Hill sheep production systems development. *Hill Farming Research Organisation 5th Report, 1966–70*: 70–87.

Evans, D.M. 1973. Seasonal variations in the body composition and nutrition of the vole *Microtus agrestis. Journal of Animal Ecology, 42*: 1–18.

Ferns, P.N. 1976. Diet of *Microtus agrestis* population in southwest Britain. *Oikos, 27*: 506–511.

Gething, P.A., Newbould, P., & Patterson, J.B.W. eds 1973. *Hill Pasture Improvement and its Economic Utilisation.* Colloquium Proceedings No. 3 of Potassium Institute Ltd, Edinburgh, 1972. Henley-on-Thames, Oxon.

Hansson, L. 1971a. Small rodent food, feeding and population dynamics. A comparison between granivorous and herbivorous species in Scandinavia. *Oikos, 22*: 183–198.

Hansson, L. 1971b. Habitat, food and population dynamics of the field vole in south Sweden. *Viltrevy, 8*: 267–378.

Hibbard, B.G. 1985. Restructuring of plantations in Kielder Forest District. *Forestry, 58*: 119–129.

Korpimäki, E. 1985. Clutch size and breeding success in relation to nest box size in Tengmalm's owl, *Aegolius funereus. Holarctic Ecology, 8*: 178–180.

Newton, I. 1979. *Population ecology of raptors.* Poyser, Berkhamstead.

Petty, S.J. 1987. Breeding of Tawny Owls *Strix aluco* in relation to their food supply in an upland forest. *In: Breeding and Management of Birds of Prey,* ed. by D.J. Hill, 167–179. University of Bristol, Bristol.

Ratcliffe, P.R. & Petty, S.J. 1986. The management of commercial forests for wildlife. *In: Trees and Wildlife in the Scottish Uplands,* ed. by D. Jenkins. Institute of Terrestrial Ecology Symposium no 17, Banchory.

Summerhayes, V.S. 1941. The effect of voles *Microtus agrestis* on vegetation. *Journal of Ecology, 29*: 14–48.

Taylor, I.R., Dowell, A., Irving, T., Langford, I.K. & Shaw, G. 1988. The distribution of the barn owl *Tyto alba* in south west Scotland. *Scottish Birds, 15*: 40–43.

Village, A. 1981. The diet and breeding of long-eared owls in relation to vole numbers. *Bird Study, 28*: 215–224.

Village, A. 1982. Home range and density of kestrels in relation to vole abundance. *Journal of Animal Ecology, 51*: 413–428.

Village, A. 1983. The role of nest site availability and territorial behaviour in limiting the breeding density of kestrels. *Journal of Animal Ecology, 52*: 635–648.

Village, A. 1987. Numbers, territory size and turnover of short-eared owls *Asio flammeus* in relation to vole abundance. *Ornis. Scandinavica, 18*: 198–204.

Genetic differentation in a released population of Eagle Owls *Bubo bubo*

K. Radler

Radler, K. 1992. Genetic differentiation in a released population of Eagle Owls Bubo bubo. *In: The ecology and conservation of European owls*, ed. by C.A Galbraith, I.R. Taylor and S. Percival, 22-27. Peterborough, Joint Nature Conservation Committee. (UK Nature Conservation, No. 5.)

As part of a reintroduction project, basic genetic data have been collected utilising a common isozyme analysis method. This paper provides empirical evidence for genotypic viability selection, at a polymorphic locus coding for a plasma esterase, by considering differential genotype mortality prior to and after release of cohorts of juveniles. With the exception of the major cause of death (electrocution), all types of mortality tended to be selective at this locus. To aid the conservation of the Eagle Owl *Bubo bubo*, two proposals are made: (a) electrocution by power poles is not selective therefore elimination of its effect should be a high priority; and (b) a captive population for a reintroduction project should be recruited from wild nestlings rather than from injured birds.

Karl Radler, Department of Forest Genetics, Georg-August-University Buesgenweg 2, D-3400 Goettingen, Germany.

Introduction

The Eagle Owl *Bubo bubo* had disappeared from most parts of the Federal Republic of Germany by the 1960s, with the exception of a small and declining population in Bavaria. During that time, attempts were made in several regions of the country to reintroduce the species into areas which still supported suitable habitats. More than a hundered birds have been released each year, since the 1980s, in this programme. These releases have resulted in the reoccupation of suitable habitats in various areas. The release of Eagle Owls was organised by a private initiative named Aktion Zur Wiederienburgerung des Uhus (AZWU). AZWU collaborated with zoos and wildlife parks to obtain owls, descendent from captive populations, for the releases. The objectives, history and status of the project are summerised in greater detail by Radler & Bergerhausen (1988) and Bergerhausen & Radler (1989).

Several years ago a study to gather emiprical data for the assessment of genetic aspects of conservation was based around the reintroduction project. This paper reports on the results from preliminary work which was undertaken to examine selective mortality.

Materials and methods

Members of AZWU collected blood samples from most of the owls used in the reintroduction project, employing the sampling method described in Radler (1986). To obtain basic genetic data standard biochemical methods of starch gel electrophoresis were used. The methods were adapted, in order to obtain optimal resolution with the sample material, from those used for analyses on human blood (Harris & Hopkinson 1976).

In a preliminary screening, two of eight enzymes revealed a polymorphism – a plasma esterase (EST) and phosphogluconate dehydrogenase (PGD) in the red cells. The first of these was used to test the hypotheses presented in this paper, after the genetic control of the observed electromorphs (phenotypes) had been confirmed by a classic Mendelian segregation analysis (Radler 1986).

Estimation of age-specific survival rates could not be made as the recovery data, due to considerable ring losses, were probably biased. In addition there were conceptual difficulties (Lakhani 1985). However, the table of releases and subsequent recoveries indicated that, within the first year of life, between 30 and 80% of the released owls died (Radler & Bergerhausen 1988). This considerable reduction within each cohort could have resulted in substantial genotypic selection.

The released population can be divided into two groups: those recovered (dead or injured) and those not recovered, which include both surviving birds and those that died but were never reported. Recovered individuals were recorded in five different mortality categories (Table 1). Of those individuals recovered the following question was asked: Is the genotypic structure of the released population differentiated among different types of mortality recorded ?

Table 1. Classification of the released population.

Abbreviation	Definition
REL	Released Population
-SUR	Survived (i.e. not recovered) proportion
-REC	Recovered (injured or dead) due to
(1) ELEC	Electric power poles
(2) TRAF	Traffic (road or railway)
(3) WIRE	Wires or fences
(4) SICK	Sickness or malnutrition
(5) REST	Other causes or missing information

Three genetic measures were utilised:

(1) The absolute distance between two frequency distributions:

$$d_o = 1/2 \sum_i |p_i - q_i|$$

where p_i and q_i are the relative frequencies of the genetic types (here genotypes) in the two populations. This measure was selected as it demonstrated the most favourable mathematical properties (Gregorius 1984).

(2) To quantify the level of differentiation among subsamples (or subpopulations) a measure based on the absolute genetic distance d_o was used (Gregorius & Roberts 1986):

$$D_j = 1/2 \sum_i |p_{ij} - p_{ij}^c|$$

The distance between the genotypic frequencies p_{ij} of each of the mortality categories (the j-th subsample) and the lumped remaining classes p_{ij}^c were measured

(3) Heterozygosity was measured as the proportion H_o of heterozygous individuals in a subsample.

Results

Differentiation due to cause of death.

Table 2 illustrates the observed genotype frequencies, or genotypic structures, as well as the calculated genetic measures. The following results were important:

(1) The differentiation D_j was lowest for the mortality factor ELEC ($D_j = 2.2\%$) and highest for the birds categorised as SICK ($D_j = 15.0\%$), whilst the mortality factor TRAF and WIRE were ranked between these extreme values ($D_j = 5.3$ and 10.3% respectively).

(2) None of the differences D_j is statistically significant. However, this is not surprising, since the sample size, even in the most differentiated mortality class (SICK), is very small.

(3) The proportion of heterozygotes was highest in the mortality factor ELEC ($H_o = 24.1\%$) and lowest in the mortality class SICK ($H_o = 9.1\%$).

The observed trends of differentiation can be explained as follows:

(a) The absence of differentiation of the mortality class ELEC was due to the random effect of this factor.

(b) The genotypic structure of the ELEC class contrasted with the distinct deviations of the genotypic frequency of the class SICK. SICK presumably reflects the 'natural' mortality.

(c) The genotypic structures of the two other mortality categories, TRAF and WIRE, showed only a moderate differentiation, since they were composed of both a random and selective portion.

(d) A fairly high percentage of the observed deaths and injuries were caused at random (see the sub-sample sizes N_j) This would explain why the genotypic structure of the subsample SUR (proportion not recovered) had such a small differentiation.

Although the observed differences D_j were not statistically significant, due to small sample sizes, the

Table 2. Genotypic frequencies (%), differentiation due to types of mortality and observed proportion of heterozygotes.

	REL	SUR	REC				
			ELEC (1)	TRAF (2)	WIRE (3)	SICK (4)	REST
N_j	558	351	79	49	23	11	45
c_j	100	62.9	14.2	8.8	4.1	2.0	8.1
EST-11	3.4	3.4	5.1	2.0	–	9.1	2.2
EST-12	23.8	25.4	24.1	20.4	17.4	9.1	22.2
EST-22	72.8	71.2	70.9	77.6	82.6	81.8	75.6
$D_j(\%)$		4.1	2.2	5.3	10.3	15.0	3.0
$H_o(\%)$	23.8	25.4	24.1	20.4	17.4	9.1	22.2

assumptions made can be explained as follows:

(1) Eagle Owls use electric power poles for perching on. The construction of modern poles, however, is such that perched owls can easily complete the circuit between the live wires and the earth wire, resulting in electrocution. Assuming that mortality caused by ELEC is random, any recoveries may be taken as a random sample, from which the genotypic structure of the survivors, in a given age class, can be estimated. This implies that if the other types of mortality are selective there would be a genotypic difference between first and second year mortalities by electrocution. Table 3 confirms this prediction. Although there was little genetic distance between the released population (REL) and the recoveries, during the first year of life ($d_o = 1.3\%$), the genetic frequency distributions differed by 29.7% and 30.0% from the recoveries of birds which were older than one year. As expected, the distance to the class SICK was even higher: $d_o = 37.4\%$. The observed increase in heterozygosity H_o (from 23.4 to 44.6) tells us that this is due to a lower mortality of the heterozygous animals.

(2) In another study (Berghausen pers comm) post mortems on a sample of recovered Eagle Owls, over a ten year period, were undertaken. Investigations to identify abnormalities in the kidneys or liver in each of the mortality classes were carried out by vets. Amongst recoveries from the class ELEC only 14% (4 out of 29) showed signs of kidney or liver abnormalities. However in the class TRAF 33% (11 out of 31) showed signs of these abnormalities and 32% (9 out of 28) in the mortality class WIRE. In the present study (see Table 2) thirteen Eagle Owls were diagnosed with kidney or liver abnormalities among the birds classified into

mortality types ELEC, TRAF or WIRE. The genotypic structure of these birds was very similar to those in the class SICK. This indicated that the thirteen Eagle Owls should be included in the subsample SICK resulting in the reclassification of individuals within mortality groups as illustrated in Table 4.

Reclassification of the individuals (Table 4) resulted in a smaller differentiation value D_j of the classes TRAF and WIRE, but a higher differentiation value of the mortality class SICK. The D_j value of 16.2% of the SICK class was nearly seven times greater than compared with the value of 2.4% of the ELEC class. Once again this higher differentiation was apparently attributed to a higher mortality of homozygote genotypes. The heterozygosity of the class SICK, the proportion of the heterozygous birds in the released population, $H_o = 8.3\%$ (2 out of 24) was much lower than the expected value; $H_o = 23.8\%$. The author tested the statistical significance of this difference by a binomial test (Sokal & Rohlf 1973), calculating $B(R_{12}, N, P_{12})$ which equals the probability of observing not more than R_{12} heterozygous birds among the N recoveries due to this type of mortality, where P_{12} is the expected proportion if random mortality is assumed. This resulted in a value of $B(2, 23, .238) = 5.2\%$, slightly above the common significance limit of 5%. As the cause of death within the mortality class SICK most likely reflects 'natural mortality', after fledging or release, the coefficient of relative viability selection against the homozygotes, was calculated, enabling an order of magnitude to be ascertained.

Viability selection

The problem with any recovery data is the unknown reporting rate therefore the coefficient of selection can only be given as a function of this rate.

If N_{ij} was the number of released and M_{ij} the number of dead birds per genotype (with $N = \Sigma N_{ij}$ and $M = \Sigma R_{ij}$), then the relative mortalities and viabilities per genotype would be:

$$m_{ij} = M_{ij} / N_{ij}$$

$$v_{ij} = (N_{ij} - M_{ij}) / N_{ij} = 1 - m_{ij}$$

and the coefficient of selection agains homozygotes at a two-allele-locus:

$$s_{ij} = (v_{ij} - v_{ij}) / v_{ij} = (m_{ii} - m_{ij}) / (1 - m_{ij})$$

In addition, if R_{ij} was the number of birds recovered per genotype (with $R = \Sigma R_{ij}$) as well as q, the proportion recovered among all dead birds (with $q = R/M < 1$) and assuming an equal reporting

Table 3. Genotypic distances between the released population, the recoveries due to electricity in the first and second year of life, and the group SICK.

	REL	ELEC 1st Year	2nd Year	SICK
Nj	410	47	9	11
c_j (%)		11.5	2.2	
EST-11	3.4	2.1	11.1	9.1
EST-12	22.4	23.4	44.4	9.1
EST-22	74.1	74.5	44.4	81.8
H_o (%)	22.4	23.4	44.4	9.1
d_o (%):				
REL		1.3	29.7	13.3
ELEC 1. Year of life			30.0	14.3
ELEC 2. Year of life				37.4

Table 4. Genotypic frequencies (%) differentiation due to types of mortality and observed proportion of heterozygotes (Table 2 after reconstruction, see text)

	REL	SUR	REC ELEC (1)	TRAF (2)	WIRE (3)	SICK (4)	REST
N_j	558	351	75	45	18	24	45
c_j (%)	100	62.9	13.4	8.1	3.2	4.3	8.1
EST-11	3.4	3.4	5.3	2.2	–	4.2	2.2
EST-12	23.8	25.4	24.0	22.2	22.2	8.3	22.2
EST-22	72.8	71.2	70.7	75.6	77.8	87.5	75.6
D_j (%)		4.1	2.4	3.0	5.2	16.2	3.0
H_o (%)	23.8	25.4	24.0	22.2	22.2	8.3	22.2

probability for each genotype the following relations result:

$$R_{ij} = qM_{ij} \text{ or } M_{ij} = R_{ij}/q$$

and

$$m_{ij} = M_{ij}/N_{ij} = \frac{R_{ij}}{qN_{ij}}$$

which results in the formula adjusted for the unkown reporting rate

$$s_{ii} = \frac{R_{ij}/N_{ii} - R_{ij}/N_{ij}}{q - R_{ij}/N_{ij}}$$

Assuming a reporting rate of $q = 100\%$, the selection against the homozygote genotypes would be $s_{ii} = 3.7\%$ whilst with a, presumably more realistic reporting rate, of $q = 50\%$ the value for selection against the homozygote genotypes would be $s_{ii} = 7.6\%$.

The method used to detect viability selection was a direct approach, known as cohort analysis. This method did not suffer from the shortcomings of other methods (see Endler 1986). The calculated values for the coefficient of selection s_{ii} were only crude estimates since they were (a) based on a very small sample size and could not be referred to a precise age class, but rather ranged between 1 and 4 years of age after release (with most emphasis on the first year), and (b) cohorts that were observed for unequal time spans were combined, due to the small sample size.

The genetic structure of each cohort was analysed approximately 12 weeks after hatching. This enabled genotpyic selection to be measured prior to release. The average mortality rate during this period was estimated at 33% (Bergerhausen et al. 1989). Since the genotype is only observable at the end of this stage, the frequency of each genotype was compared with its expected frequency according to Mendelian laws. Matings of heterozygotes with a homozygote

genotype (EST-12 × EST-22) resulted in 186 offspring with the following ratio:

EST-12:EST-22 = 101:85 = 54%:46%

whilst the expected ratio would have been

EST-12:EST-22 = 50%:50% = 93:93

From this an estimate of coefficient of viability, for the offspring genotypes of $v_{12} = 101/93 = 1.086$ and $v_{22} = 85/93 = 0.914$ was made, resulting in a selective disadvantage of the homozygote

$$s_{22} = \frac{(v_{12} - v_{22})}{v_{12}} = \frac{1.086 - 0.914}{0.914} = 0.158$$

This value was two times greater than that estimated for selection amongst released owls, as shown above. However, given the sample size, the selection difference was not large enough to be statistically significant. Provided therefore, that a larger sample is not yet available, the results presented can be considered as further indication of genotypic selection among zygotes (and/or possibly gametes prior to fertilization).

Discussion

Heterozygosity at enzyme loci has been indicated as a selective advantage for several species (see Mitton & Grant 1984). The evidence for genotypic selective mortality, as described above, prior to and after release of juvenile Eagle Owls should therefore not be surprising. However, interpretation of genotypic selection at one single enzyme gene locus must be treated with caution for the following reasons:

(1) Viability selection of this order of magnitude due to one single gene locus appears high, as viability is affected by several loci. However, endangered species, throughout a number of successive generations have low population sizes and therefore a reduced genetic diversity due to genetic drift. As a consequence of this, the

25

genetic background (of the locus considered here) would tend to be less diverse, making compensating effects of other polymorphic loci (e.g. on viability) less likely.

(2) The sample also contains an unknown number of inbred birds. If these were not distributed at random across the three genotypes, inbreeding depression, due to sublethal recessive alleles at other loci, could have caused the viability differences noted. However, there are two counter arguments to this: (a) inbreeding depression affects viability mainly in an early stage of the life cycle (Templeton & Read 1984) and not (or very little) beyond the age of 12 weeks, when the owls in this programme were released, and (b) the same directional trend, that of selection against homozygotes within genetically segregating broods, was observed.

(3) There is also the possibility, that the selection is caused by closely linked loci rather than the esterase itself. However, there are two counter arguments to this theory:

(a) Basic population genetic theory suggests an adaptive meaning for any major polymorphism (Lewontin 1986) – like the esterase in my case, especially in a K-selected species with typically small populations, like the Eagle Owl, because this reduces the probability of fixation due to genetic drift as compared to a neutral locus.

(b) My second argument is physiological and entirely speculative as yet, for no experimental work has been done. Basic biochemical knowledge could suggest esterases decompose ester-bonds, and it is possible that the two occurring esterase isozymes differ in their molecular turnover number with respect to substrate or physiological environment, as it has been demonstrated in vitro with a fish species by Koehn (1969). If our esterase is operating in a similar way in Eagle Owls, this should give an advantage to a heterozygous individual, because it might be more effective in transforming ester bonds – either taken up with its food or stored in various fat deposits – into metobolic energy. The evolutionary advantage of that is obvious, if one recalls three properties of the ecology of this species (see e.g. Glutz H von Blotzheim & Bauer 1980): (a) the Eagle Owl is a typical generalist with respect to its diet, (b) it is a long-living and repeatedly (iteroparic) reproducing vertebrate, and (c) in our climate zones – and especially after the alterations in its habitat – this bird frequently experiences critical periods of food shortage due to unfavourable seasons and weather conditions.

Conservation implications

(1) The most frequent cause of Eagle Owl death is electrocution by power poles. The available demographic data suggests that the mortality level resultant from this factor is crucial to the survival of this reintroduced population (Radler & Bergerhausen 1988 and unpublished; and Larsen *et al.* 1987). Viability selection could be an effective means to adapt a population's genetic structure to a changed or new environment, which has particular relevance for a reintroduced population. In this context the results presented in this paper should be viewed as an empirical demonstration that some causes of death by traffic, wires and fences, sickness and malnutrition tend to select in a 'natural' or adaptive way, whilst the most frequent cause of death, due to electrocution, does not result in selective adaptation in reintroduced populations. In addition the level of mortality by electrocution prevents the released population from growing at a rate which is sufficient to minimise the loss of genetic diversity by genetic drift. The successful conservation of Eagle Owl populations should therefore be concentrated on modification of the construction build of power poles in such a way that the risk of electrocution is minimised.

(2) The AZWU programme frequently recruited the captive population from injured wild birds. This, however, is not an optimal strategy since these birds are likely to be less fit; their offspring are likely to have lower chances of survival. Captive breeding from injured wild birds is laborious and costly. Therefore, if a captive population for a re-stocking project is to be established, wild birds should be favoured provided that the source population is not in decline. The most appropriate source of birds are nestlings of adult pairs which have an above average fecundity. Management of the population in this way would tend to equalize the individual fitness (lifetime repoductive success), and could even bring about an increase of genetic effective population size (Crawford 1984) in the source population. Management which focuses on the preservation of the genetic adaptability is a desirable practice of modern conservation strategies (Frankel & Soule 1981).

26

Acknowledgements

Publication of this work was rendered possible through funds provide by the Bundesministerium fuer Umwelt. Naturschutz and Reaktorischerheit of the Federal Republic of Germany. I am also much indebted to all the contributing members, donors and field ornithologists of the 'Aktion zur Wiedereinbuergerung des Uhus (AZWU)' for various kinds of support, and I would especially like to thank W. Bergerhausen, the coordinator of the AZWU, for his continuous and very valuable cooperation, and all my colleagues for helpful comments and valuable discussions at our department.

This paper has benefited from comments (including linguistic improvements) of Laurel Hanna as well as the editors of this book, and an anonymous reviewer, which are gratefully acknowledged.

References

Bergerhausen, W. & Radler, K. 1989, Bilanz der Wiedereinbuergerung des Uhus (*Bubo bubo* L.) in der Bundersrepublik Deutschland. *Natur und Landschaft, 64*: 151–161.

Bergerhausen, W., Radler, K. & Willems, H. 1989. Reproduktion des Uhus (*Bubo bubo* L) in verschiedenen europaeischen. Teilpopulationen sowie einer Population in Gehegen. *Charadrius. 2*, 85–93.

Crawford, T.J. 1984. What is a population ? *In: Evolutionary ecology*, ed. by B. Shorrocks, 135–173 Oxford, Blackwell.

Endler, J.A. 1986. *Natural selection in the wild.* Princeton & New York, University Press.

Frankel, O.H. & Soule, M.E. 1981. *Conservation and evolution.* Cambridge, University Press.

Glutz H von Blotzheim, & Bauer, K.M. 1980. *Handbuch der Vogel Mitteleuropas.* Bd 9 (*Columbiformes-Piciformes*). Wiesbaden, Akad. Verlagsgesellschaft.

Gregorius, H.R. 1984. A unique genetic distance. *Biometrics Journal, 26*: 13–18.

Gregorius, H.R. & Roberds, J.H. 1986. Measurement of genetical differentiation among subpopulations. *Theoretical and Applied Genetics, 71*: 826–834.

Harris, H. & Hopkinson, D.A. 1976. *Handbook of enzyme electrophoresis in human genetics.* Amsterdam, Elsevier-North Holland.

Koehn, R.K. 1969. Esterase heterogeneity: dynamics of a polymorphism. *Science, 163*: 943–944.

Lakhani, K.H. 1985. Inherent difficulties in estimating age-specific bird survival rates from ring recoveries. *In: Statistics in ornithology*, ed. by Morgan & North. 311–321 Berlin, Springer.

Larsen, R.S., Sonerud, G.A., & Stensrud, O.H. 1987. Dispersal and mortality of juvenile Eagle Owls released from captivity in southeast Norway as revealed by radio telemetry. *In: Biology and conservation of northern forest owls. Sysmposium proceedings*, ed. by Nero *et al.* 215–219, Winnipeg, Manitoba.

Lewontin, R.C. 1986. Population genetics. *Annual Review of Genetics, 19*: 81–102.

Mitton, J.B. & Grant, M.C. 1984. Association among protein heterozygosity, growth rate, and development homeostasis. *Annual Review of Ecology and Systematics, 15*: 479–499.

Radler, K. 1986. Polymorphism and genetic control of a plasma esterase in the Eagle Owl. *Heredity, 56*: 65–67.

Radler, K. & Bergerhausen, W. 1988. On the life history of a reintroduced population of Eagle Owls (*Bubo bubo*). *In: Proceedings of the international symposium on raptor reintroduction 1985*, ed. by D.G. Garcelon & G.W. Roemer, 83–94. Arcata, Institute for Wildlife Studies.

Sokal, R.R. & Rohlf, F.J. 1973. *Introduction to biostatistics.* San Francisco, Freeman and Co.

Templeton, A.R. & Read, B. 1984. Factors eliminating inbreeding depression in a captive herd of Speke's gazelle (*Gazella spekei*). *Zoo Biology, 3*: 177–199.

Population studies of the Ural Owl *Strix uralensis* in Finland

P. Saurola

Saurola, P. 1992. Population studies of the Ural Owl *Strix uralensis* in Finland. *In: The ecology and conservation of European owls*, ed. by C.A. Galbraith, I.R.Taylor and S. Percival, 28-31. Peterborough, Joint Nature Conservation Committee. (UK Nature Conservation, No. 5.)

The Ural Owl *Strix uralensis*, a medium-sized bird of prey breeding in northern forests of the Palearctic, has been, together with other owls, an attractive target for Finnish bird ringers during the last 25 years. In the early 1960s the Ural Owl was considered a rare species, which suffered heavily from a shortage of natural nest-sites in old hollow trees and chimney-like stumps, these having been eliminated by short-sighted forestry practices. For this reason, an attempt to compensate losses of natural nest-sites with nestboxes was made by some ringers in southern Finland, around the city of Hämeenlinna. The operation was successful, nestboxes were accepted by the owls, and more ringers started similar activities in other parts of Finland. In addition to conservation aim, the studies have produced relevant data on the biology of hole-nesting owls, and much has already been analysed. This paper is a short review of studies on the population ecology of Ural Owls in Finland.

Pertti Saurola, Zoological Museum, P. Rautatiek. 13, SF-00100 Helsinki, Finland.

Introduction

Data on owls have been collected annually by Finnish ringers: e.g. in 1989, 1026 Ural Owl *Strix uralensis* territories and 818 nests were found, 2002 nestlings ringed and 604 adults ringed or retrapped at the nest (Table 1). In the late 1960s, only nestlings and a few brooding females were ringed at the nests found. In the 1970s and 1980s, most of the breeding females were ringed or retrapped, but up till now, only a small percentage of the breeding males have been caught systematically (for method, see Saurola 1987).

The population biology of the Ural Owl has been studied most intensively in two study areas:

1. In "Kanta-Häme" (around Hämeenlinna, 61° 00' N, 24° 30' E, 3,000 km²) 1643 active nests have been found, almost 3000 nestlings ringed and 443 different females and 219 males recorded as breeding birds during 1965–1988 (Pietiäinen, Saurola & Kolunen 1984; Pietiäinen, Saurola & Väisänen 1986; Saurola 1987, 1989a, 1989b).

2. In "Päijät-Häme" (around Lahti 61° 00' N, 25° 40' E, 1,500 km²) 477 active nests have been found during 1977–1988 (Pietiäinen, Saurola & Kolunen, 1984; Pietiäinen, Saurola & Väisänen 1986; Pietiäinen 1988, 1989). In both of these long-term projects, moult records and biometrical data on eggs, nestlings and adults have been gathered in addition to traditional data on breeding performance.

Annual production

The Ural Owl is a generalist feeder, but successful reproduction is highly dependent on fluctuating small mammal populations (Linkola & Myllymäki 1969; Korpimäki & Sulkava 1987). In Kanta-Häme, during 1965–1989, the median laying date was 6 April, but in years of peak microtine vole populations egg laying started two weeks earlier, and in low microtine years two weeks later than the overall median value. Furthermore, in springs of low vole populations, up to 90% of pairs did not lay at all and the average clutch size (range from 1 to 8) was 2.0 compared with 4.0 in high vole years. On average, 17% of active nests failed totally. The average annual production of young per breeding attempt varied from 0.9 (1981) to 2.9 (1973) and the total number of fledglings produced per year in that same area and period varied from 12 (1971 and 1981) to 336 (1973).

The Finnish Ringing Centre has stored all ringing data (not only recoveries) on a computer file since 1974 (data from larger birds since 1973). These data indicate clearly (although the ringing activity has increased in many areas) that the annual productivity of Ural Owls in 1973–1989 was synchronous over a wide area in southern Finland. A clear three-year cycle (two good years, one bad year) was detected in the 1980s but not in the 1970s (Figure 1).

Dispersal

No statistical difference in natal dispersal between male and female Ural Owls has been found so far; 50% of both sexes (72 females, 63 males) ringed as nestlings in Kanta-Häme during the last ten years, and retrapped as breeders at the nest, moved less than 20 km from their birth places. The longest distances of natal dispersal have been around 100 km for both sexes.

Table 1.   Number of territories and nests found, and nestlings and full-grown birds ringed of different owl species in Finland in the Peak Vole Year 1989. Note: Adults ringed outside the breeding season included.

	Territ. found	Nests found	Pulli ringed	Adults ringed
Eagle Owl Bubo *bubo*	996	494	854	18
Hawk Owk *Surnia ulula*	143	103	399	47
Pygmy Owl *Glaucidium passerinum*	282	81	212	89
Tawny Owl *Strix aluco*	591	420	1,168	135
Ural Owl *Strix uralensis*	1,026	818	2,004	202
Great Grey Owl *Strix nebulosa*	127	96	200	14
Long-eared Owl *Asio otus*	717	519	395	252
Short-eared Owl *Asio flammeus*	280	99	156	70
Tengmalm's Owl *Aegolius funereus*	3,637	2,259	6,678	3,526

In southern Finland, Ural Owls were more faithful to their previous nest-sites than Tawny Owls *Strix aluco* (Saurola 1987). So far, no Ural Owl males and only 4% of females have been retrapped and 15% found dead more than 5 km away from their previous nest (corresponding figures were 3% and 24% for the male and 10% and 24% for the female Tawny Owls). A small proportion of the Ural Owl females left their territories but only in very harsh winters and, if some of these few emigrants survived, they may have bred quite far from their previous nest-sites. In general, most of the Ural Owls remained in their territories during poor breeding seasons as non-breeding birds.

The high probability of Ural Owls breeding with the same mate year after year may be a direct consequence of the high site tenacity of both sexes: the estimated "divorce rate" in the Ural Owl was 3–5% (15–20% in the Tawny Owl, Saurola 1987).

Survival

A sophisticated analysis of the age- and time-specific survival rates in the Ural Owl is still in preparation. Preliminary estimates for annual survival, based on retraps and random recoveries, are 60% during the first, 70% during the second, 80% during the third and 85% during the fourth and later years of life (Saurola 1989b). The longevity record for the Ural Owl is 21 years for a female which was ringed as a breeding bird in 1969 and was retrapped at the nest in 1989.

Age at first breeding

Of a sample of 82 females, which were all ringed as nestlings eleven years before the analysis was made, 59% started to breed at the age of three or four years, 12% started earlier (1–2 years) and 29% later (5–9 years) (Saurola 1989b). By omitting recent data, late starters will be correctly represented in the analysis. The age distribution of breeding females is highly dependent on the quality of the year: in poor vole years, all breeders were at least 5 years old, whilst in good years some of the one-year-old females started to breed (Saurola 1989a). However, starting at the age of one year may be a risk for survival, because the number of flight feathers which can be moulted is dependent on the number of young produced (Pietiäinen, Saurola & Kolunen 1984). It is particularly important for one-year-old birds to change as many as possible of the low quality juvenile feathers before their second winter (Saurola 1989a).

Recruitment

Recruitment rate of fledglings is dependent on survival from fledging to the first breeding and on the number of free territories available for potential recruits. Recruitment of Ural Owl fledglings from Kanta-Häme varied greatly from year to year, but was not as clearly related to the prey abundance (Saurola 1989b) as suggested by Korpimäki & Lagerstrom (1988) for the Tengmalm's Owl *Aegolius funereus*. In the 1980s recruitment rate was very low for fledglings hatched during a peak just before a poor microtine year (cohorts 1980, 1983, 1986; Figures 2 and 3), but in the 1970s the relationship was obscure. Because of differences in production and recruitment rates, different cohorts were unequally represented in the breeding population. Only very rarely are production, survival of fledglings and availability of territories all favourable for the same cohort. During the last two decades, year-class 1973 (a peak year just before a low microtine) year was such an exceptionally 'lucky' cohort (Figures 1 and 2).

Lifetime Reproduction

Lifetime reproductive success of Ural Owl females has been analysed on the basis of data from Kanta-

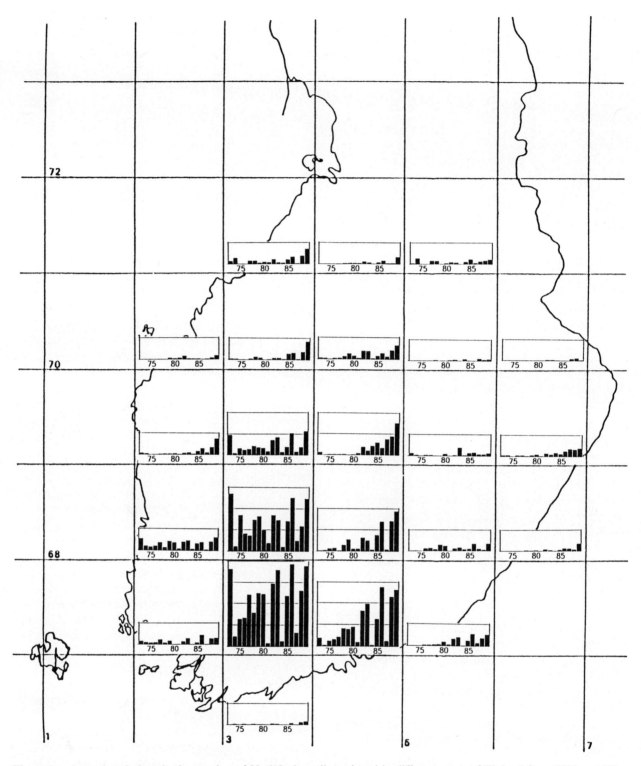

Figure 1. Annual variations in the number of Ural Owl nestlings ringed in different parts of Finland from 1973 to 1989 (100 km × 100 km squares of National Grid, vertical scale = 100 nestlings).

Häme (Saurola 1989b). The maximum lifespan of this species is so long that even these data, after 25 years of work, are not yet representative enough for a perfect analysis. Estimates of lifetime reproduction which are based on a much shorter study period than the normal lifespan of the species (Pietiäinen 1989) are of limited value. According to my data, lifetime reproductive output of Ural Owl females was highly variable: 50% of fledglings were produced by 23% of breeding females or 6% of fledglings from the previous generation. The most variable components of lifetime production were the recruitment rate of offspring, lifespan and the number of breeding attempts per year. Successful females had, on

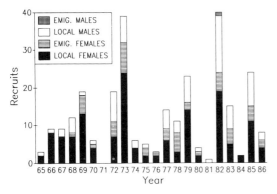

Figure 2. Numbers of verified recruits from fledglings ringed in Kanta-Häme study area during 1965–1986. Note: (a) the study population was still growing during the last years of the 1960s; (b) the effort of catching females at the nest has been on the same level since 1970 in Kanta-Häme ('local females') and since the late 1970s in surrounding areas ('emigrant females'); and (c) the effort of catching males has been on the same level only in my own study area (a part of Kanta-Häme, Saurola 1989b) since 1976, hence, the data of male recruits from the 1970s and 1980s are not comparable.

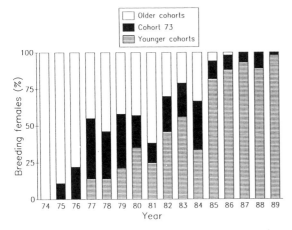

Figure 3. Number of female recruits (%) from cohort 1973 among breeding females in Kanta-Häme in 1974–89.

average, longer wings and started to breed earlier, but were not more aggressive in defending their nests than the less successful ones.

Future of the Finnish population

The intensively studied breeding population in Kanta-Häme has fluctuated within the same limits (9–120 active nests per year) through the last two decades and, in many other parts of Finland, the number of Ural Owl nests in nestboxes is still increasing (Figure 1). Thus, at the moment, the prediction for the Finnish Ural Owl population is positive, but only if amateur ornithologists continue to compensate with appropriate nestboxes for nest-

site losses caused by forestry. In many areas no natural nest-sites remain that are good enough for normal reproduction. In future, a rapidly increasing Eagle Owl *Bubo bubo* population will be another factor, dependent on human activities (rubbish dumps and clear cuts), which may restrict at least locally the limits of the Finnish Ural Owl population.

Additional remarks

In summer 1989, a man walking in a forest in southern Finland was suddenly attacked by an Ural Owl. The man lost his eye as a result of this attack. Damage by elks to farmers and car drivers is compensated by the state, but no compensation was paid for the injury caused by the owl. For this reason, it is very important that the locations of all new owl nestboxes on private land have been agreed with the landowner and he is made aware of the risks. In Finland, very few ringers have so far asked permission for such direct conservation action such as constructing nestboxes or other artificial nests for birds of prey.

References

Haapala, J., Korhonen, J., & Saurola, P. 1990. Breeding of raptors and owls in Finland in 1989 *Lintumies, 25*: 2–10. (In Finnish with English summary).

Korpimäki, E., & Lagerström, M. 1988. Survival and natal dispersal of fledglings of Tengmalm's Owl in relation to fluctuating food conditions and hatching date. *Journal of Animal Ecology, 57*: 433–441.

Korpimäki, E., & Sulkava, S. 1987. Diet and breeding performance of Ural Owls under fluctuating food conditions. *Ornis Fennica, 64*: 57–66.

Linkola, P., & Myllymäki, A. 1969. Der Einfluss der Kleinsäugerfluktuationen auf das Brüten einiger Kleinsäugerfressender Vögel in südlichen Häme, Mittelfinnland 1952–1966. *Ornis Fennica, 46*: 45–78.

Pietiäinen, H. 1988. Breeding season quality, age and the effect of experience on the reproductive success of Ural Owls, *Strix uralensis. Auk, 105*: 316–324.

Pietiäinen, H. 1989. Seasonal and individual variation in the production of offspring in the Ural Owl *Strix uralensis. Journal of Animal Ecology, 58*: 905–920.

Pietiäinen, H., Saurola, P. & Väisänen, R. 1986. Parental investment in clutch size and egg size in the Ural Owl *Strix uralensis. Ornis Scandinavica, 17*: 309–325.

Pietiäinen, H., Saurola, P., & Kolunen, H. 1984. The reproductive constraints on moult in the Ural Owl *Strix uralensis. Ornis Fennica, 17*: 309–325.

Saurola, P. 1987. Mate and nest-site fidelity in Ural and Tawny Owls. In: *Biology and conservation of northern forest owls: symposium proceedings, February 3–7, 1987, Winnipeg, Manitoba,* ed. by R.W. Nero., R.J. Clark, R.J. Knapton & R.H. Hamre. USDA Forest Service. General Technical Report. RM-142.

Saurola, P. 1989a. Breeding strategy of the Ural Owl *Strix uralensis. In: Raptors in the modern world,* ed. by B.U. Meyburg, & R.D. Chancellor. 235–240. Berlin, World Working Group on Birds of Prey and Owls.

Saurola, P. 1989b. Ural Owl. In: *Lifetime reproduction in birds.* ed. by I. Newton. 327–345. London, Academic Press.

An assessment of the significance of annual variations in snow cover in determining short-term population changes in Field Voles *Microtus agrestis* and Barn Owls *Tyto alba* in Britain

I.R. Taylor

Taylor, I.R. 1992. An assessment of the significance of annual variations in snow cover in determining short-term population changes in Field Voles *Microtis agrestis* and Barn Owls *Tyto alba* in Britain. *In: The ecology and conservation of European owls*, ed. by C.A. Galbraith, I.R.Taylor and S. Percival, 32-38. Peterborough, Joint Nature Conservation Committee. (UK Nature Conservation, No. 5.)

Populations of Field Voles *Microtus agrestis* were assessed by trapping each year from 1975 to 1989 in a study area in south Scotland. The number of pairs of breeding Barn Owls *Tyto alba* in the same area was determined from 1979 to 1989. Barn Owls were highly dependent on Field Voles as their main prey species.

The vole populations were cyclic with peak and decline years synchronised throughout the study area. No relationship was evident between vole population declines and variations in the duration of winter snow cover. There was also no relationship between yearly changes in Barn Owl numbers and variations in snow cover. Changes in Barn Owl numbers were significantly correlated with changes in Field Vole abundance.

The effects of the exceptionally severe winter of 1978–79 on Barn Owl numbers in south Scotland are examined. From a sample of nest sites, Barn Owl breeding numbers between sea level and 150 m a.s.l. (the altitude range within which most British Barn Owls occur) declined by 25.5% from 1978 to 1979, but numbers recovered to 96% of the 1978 level by 1981.

The significance of snow duration to long term changes in Barn Owl numbers in Britain is discussed and it is concluded that there is little evidence to support suggestions that changes in winter snow cover have been important in the long term decline of Barn Owls.

I. R. Taylor, Institute of Cell, Animal and Population Biology, University of Edinburgh, Zoology Building, West Mains Road, Edinburgh EH9 3JU

Introduction

Short-term changes in the density of breeding Barn Owls *Tyto alba* have been demonstrated in a number of European populations (Schonfeld *et al.* 1977; Kaus 1977; De Bruijn 1979; Braaksma 1980; Taylor *et al.* 1992). Such changes are usually cyclic and correlate with changes in the abundance of microtine voles (mostly the Common Vole *Microtus arvalis* and Field Vole *M. agrestis*) which are the owls' main food supply.

Vole populations have been studied extensively in Europe and North America and both cyclic and non cyclic populations have been identified. Cycles are normally of 3–4 years periodicity and attempts to elucidate their causation have concentrated on intrinsic factors, including changes in social behaviour and extrinsic factors such as food supply, cover and predation (Pearson 1966; Tast & Kalela 1971; Hansson 1979; Taitt & Krebs 1981; Hestbeck 1982; Batzli 1983; Erlinge *et al.* 1983; Haukioja *et al.* 1983; Laine & Henttonen 1983; Taitt & Krebs 1983; Hansson 1984; Henttonen 1985; Krebs 1985; Hornfeldt *et al.* 1986; Hansson 1987; Jarvinen 1987; Lindstrom *et al.* 1987).

Analyses of long-term data have shown no consistent simple relationships between annual weather patterns and population cycles (Maerks 1954; Krebs & Myers 1974; Mihok *et al.* 1985).

Shawyer (1987) attempted to investigate short-term changes in Field Vole and Barn Owl populations in Britain. For vole populations, information from local studies was extracted from the literature and for owl populations, the annual ringing totals recorded with the British Trust for Ornithology were used. The numbers of young Barn Owls ringed in the whole of Britain were totalled and expressed as a percentage of the total young of all species ringed. Changes in the index so produced were assumed to be equivalent to changes in the density of breeding Barn Owls. The index and information on vole populations were then compared with data on snow cover averaged over a number of weather stations (locations not specified) and from this it was concluded that snow duration was one of the main factors responsible for cyclic changes in Barn Owl numbers by regulating the abundance and availability of Field Voles.

The purpose of this paper is to test this suggestion by examining data on Field Vole abundance, the numbers of breeding Barn Owls and snow duration from a long-term detailed population study in south Scotland.

Methods

The study area comprised the catchment of the Esk and Liddle rivers which flow into the Solway Firth north of Carlisle, Cumbria (2° 55′ W, 54° 54′ N). The area covered approximately 1600 km² of mixed farmland and forestry rising from sea level to an altitude of 500 m. Above 150 m most of the area was sheepwalk and coniferous plantations mainly planted from the mid-1960s onwards but some dating back to the 1940s. At lower altitudes land use was mostly pastoral, for dairy and sheep production, interspersed with numerous small woodlands, again mainly of coniferous species.

Breeding Barn Owls were located each year by searching all potentially suitable nesting places and other methods including listening for calling birds. Responses to broadcast tape recorded calls were tried but found to be unreliable and too time consuming. At the start of the study all farms and derelict or disused buildings within the area were searched for signs of Barn Owls and for structures within them suitable for nesting. Large hardwood trees that might have contained suitably sized holes for nesting were not abundant and mostly confined to river courses and around farm buildings, with a few isolated individuals on hillsides and along field boundaries. At the start of the study, during the first two winters, these were systematically searched for suitable holes by teams of volunteers. This revealed very few holes of adequate size that were not waterlogged in winter and spring.

All potential nest sites were examined up to three times between March and September each year. This was essential to be sure of locating early failed breeders and late breeders in addition to those within the main laying period of April to May. Each round of searching took between 10 and 20 days fieldwork.

In addition, farmers, gamekeepers, rangers, estate and forestry workers were requested to note sightings and signs of Barn Owls. Particularly interested farmers acted as local foci by passing on information obtained in discussion with others in their neighbourhood. Two local ornithologists, well known in their communities, also acted as contacts. As the whole study area was traversed several times each year it was possible to note changes in the use of

farm buildings and add to the list of suitable nest sites as appropriate.

Conditions in the winter of 1978/79 were the most severe recorded since 1962/63. Such climatic extremes can often be instructive in demonstrating the extent of species' responses. Data on vole populations within the study area were collected over this winter (see below) but the long term study of the owl populations did not start until 1979. However, a large sample of Barn Owl nests was being examined in Dumfriesshire, within and around the study area, incidentally during other studies by professional ecologists and by amateur ornithologists who ringed the young. From these, data on the occupancy of 80 nest sites were obtained for 1978 and 1979, allowing an assessment of population changes over winter. In south Scotland, Barn Owls were about 96% faithful to their nest sites from year to year, so changes in site occupancy should have been an accurate indicator of changes in the number of breeding pairs (Taylor in prep.).

Research into Field Vole populations in the study area were begun by Charles (1981). He developed methodology based on trapping from which an annual index of spring (April) vole abundance could be calculated and has kindly allowed me to use his unpublished data covering 1975 to 1979. In the years following, the trapping programme was continued by the author, I. K. Langford and A. Chaudhry.

The method involved a grid of 24 trapping stations selected randomly with two treadle operated snap traps at each, set in sample sites of two to four year conifer plantations. Each year six sites spread throughout the study area were sampled. In each, trapping was carried out continuously over five nights with traps examined and reset every day. The number of voles caught were totalled and the mean of such totals for the six sites were taken as the annual index of abundance.

Charles (1981 and unpublished) showed that changes in vole numbers occurred synchronously over large areas and in different habitats. A good general idea of the relative abundance of voles can be obtained from observations of the density of occupied runs. Major changes in abundance are readily evident. From 1980 to 1989 observations were made of the abundance of vole signs throughout the study area, in addition to the evidence from different trapping stations, as a check on synchrony. Although not quantitative this could reliably identify peak years and years of decline.

Data on snow cover were obtained from two meteorological stations, Eskdalemuir and Carlisle, situated close to the highest and lowest points of the study area. Their mean altitude (134 m a.s.l.) was close to the median altitude of Barn Owl nesting attempts. Values of snow cover used in the analysis were means from these stations, expressed as "snow days", that is, the number of days when snow covered at least 50% of the ground surface at 0930 hrs GMT, totalled from the start of November to the end of March each winter.

The validity of using means from these stations was tested further. Data on snow cover were obtained from the Meteorological Office, Edinburgh, for eight recording stations at altitudes between 15 and 387 m a.s.l in south Scotland over the 30 year period 1959–1989. These were Palmure, Glenlochar, Glenlee, Threave, Bargrennan, Penwhirn, Bowhill, Stanhope Farm, Eskdalemuir and Leadhills. Mean duration of snow cover was linearly correlated with altitude (number of snow days = 8.6 × 0.09 m a.s.l, r = 0.95. $p < 0.001$). From this, the predicted long term value for the mean altitude of Eskdalemuir and Carlisle was 20.9 days and the actual value obtained from these stations was 21.9 days. Thus mean values were considered valid indices of snow cover for the study area.

Results

Snow cover and Field Vole populations

Field Vole populations reached peak abundances in 1975, 1981, 1984 and 1987. Major declines occurred over the winters of 1975/76, 1981/82, 1984/85 and 1987/88 (Figure 1). The peaks and declines occurred synchronously throughout the study area irrespective of altitude or habitat. Only one exception to this was noted; in 1988 there was a widespread decline but numbers remained high in an area of hill margin land measuring about 7 × 4 km, declining there the following year. From 1981 onwards the population

changes conformed to a three year periodicity but there was a six year interval between the peaks of 1975 and 1981.

No consistent relationship was evident between snow cover and changes in Field Vole numbers. The 1976 and 1988 population declines occurred with exceptionally mild winters of only 8 and 11 snow days respectively, well below the long term average of 22 days. The decline of 1982 followed a winter of above average snow cover and that of 1985, a winter of close to average. The winter of 1978/79 was the coldest of the study period with twice the average number of snow days (see later) but vole populations were at a higher density in the spring of 1979 than 1978. Overall, there was no relationship between the percentage change in the index of vole abundance between springs and the duration of snow cover in the intervening winters (r = 0.15, n = 14, n.s).

Snow cover and Barn Owl populations

From 1979 to 1989 the number of pairs of breeding owls varied cyclically with peaks in 1981, 1984 and 1987. In each of the three cycles, the decline in the number of owl pairs took place over two years with the greatest decrease in the second year. Thus, in the decline phase, Barn Owls reached their lowest abundance one year after the decline in vole abundance. Over the three cycles, yearly percentage changes in the number of pairs of breeding owls were not related to the duration of winter snow cover (r = 0.02, Figure 2). They were, however, closely correlated with changes in the abundance of Field Voles (r = 0.69, $p < 0.05$, Figure 3).

Effects of 1978/79 winter

The winter of 1978/79 was particularly severe with extremely low temperatures and prolonged deep snow

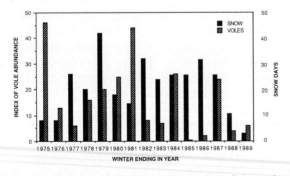

Figure 1. Relationship between the number of days of winter snow cover (November-March) recorded between 1975 and 1989 and the spring abundance of Field Voles *Microtus agrestis* in the Esk study area, south Scotland.

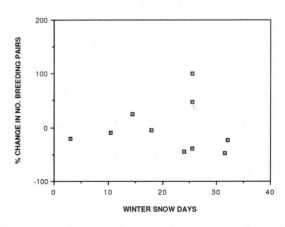

Figure 2. Relationship between the percentage change in the number of breeding pairs of Barn Owls *Tyto alba* between years and the number of days of snow in the intervening winter (November-March) in the Esk study area, south Scotland.

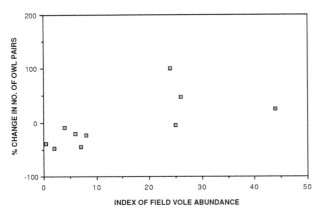

Figure 3. Relationship between yearly changes in the number of pairs of breeding Barn Owls *Tyto alba* and the spring abundance of Field Voles *Microtus agrestis* in the Esk study area, south Scotland.

cover. Mean snow duration in the study area was 42 days, twice the long term average, with 24 days recorded at Carlisle and 62 days at Eskdalemuir. Sixty-five percent of this occurred in January and February with prolonged periods of continuous snow cover during these months. Mean daily temperatures at Eskdalemuir were $-2.1°C$ for January and $-0.5°C$ for February, 3.5 and $2.0°C$ lower than the long term averages.

In 1978, breeding Barn Owls were recorded at 80 sites between 20 and 300 m a.s.l within the study area and adjacent valleys of Dumfriesshire. In 1979 breeding was recorded at 48 of these. At the remaining 32, there were no birds or recent signs at 15 sites and only single birds at 17. There was thus an overall reduction of 40% in the breeding population.

Losses were highly correlated with altitude (Figure 4). Considering only sites between 20 and

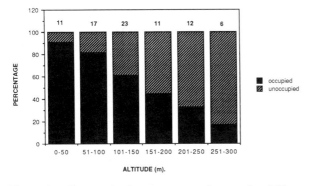

Figure 4. Changes in the occupancy of a sample of 80 nest sites by breeding Barn Owls *Tyto alba* between 1978 and 1979 in Dumfriesshire, south Scotland, in relation to altitude (m.a.s.l.). Sample sizes shown above columns. The winter of 1978/79 was the most severe recorded since 1962/63.

150 m a.s.l, the range within which most British Barn Owls nest (Bunn *et al*. 1982), breeding occurred at 51 sites in 1978 and at 38 in 1979, a decrease of only 25.5%. Single birds were present at eight sites. By 1981 breeding pairs were re-established at 49 (96%) sites. Thus, at these altitudes, the effect of this exceptionally cold and snowy winter on the number of pairs was relatively minor and short-lived. At altitudes between 150 and 300 m a.s.l the effect was more severe but, again, short-lived. Of 29 sites occupied in 1978, breeding occurred at only ten in 1979, a fall of 65.5%, with single birds at nine sites. By 1981, breeding was re-established at 25 (86%) sites.

Discussion

It is not the purpose of this paper to provide a comprehensive review of the literature pertaining to the causation of small mammal population cycles or to those of their predators. The objective was to examine the suggestion made by Shawyer (1987), that cycles of Field Voles and hence Barn Owls were caused mainly by cyclic variations in the duration of winter snow cover. Shawyer was quite precise in stating that winters with 20 or more days of snow cover were immediately followed by a decline in vole numbers and a "crash" in the Barn Owl population. Based on this, Shawyer argued that change in the long-term pattern of snow cover over Britain was a principal factor in causing long-term changes in Barn Owl numbers. Thus it is important to test Shawyer's suggestion, not only for the understanding of short-term fluctuations, but also of long-term trends and it is particularly important for the conservation of Barn Owls that the factors responsible for population change be correctly identified.

This study has shown that in south Scotland, Field Vole cycles, including the decline phases, were independent of the duration of winter snow cover. The significance of weather as a causal factor of microtine cycles has received attention from many previous workers. The synchrony that is sometimes apparent over large areas led some early researchers to suggest that weather could be involved. However, in a exhaustive review of the literature, Krebs & Myers (1974) concluded that cycles could not be simply explained by weather effects and that other factors must be involved. Chitty (1952) and Chitty & Chitty (1962) observed that populations of Field Voles within a few kilometres of each other and, therefore, experiencing similar weather, cycled out of synchrony and that declines were not associated with severe winter conditions. Geographically close populations cycling out of synchrony have been reported in a number of other studies (Middleton 1930, 1931; Tast & Kalela 1971; Schnider 1972; Myllimaki *et al.* 1977; Hanson 1979; Tapper 1979).

Long-term studies have failed to show a relationship between winter weather and vole cycles. Maerks (1954) demonstrated a three year periodicity in a 39 year series of data on Common Vole *M. arvalis* populations which could not be accounted for by variations in any weather variable. Middleton (1930, 1931) provided evidence for 3–4 year cycles of *M. agrestis* in Britain between 1900 and 1927, a period during which relatively mild, snow-free winters were prevalent. Winter and spring temperatures and snow cover were not significantly different in a year of major decline of a population of Meadow Vole *M. pennsylvanicus* compared with years when the voles were at a similar density but where no decline occurred (Mihok *et al.* 1985).

Krebs & Myers (1974) pointed out a severe conceptual difficulty with hypotheses based on weather factors, namely that cycles occur in species living in a wide variety of climates including areas with prolonged deep snow and also areas where snow is rare. In the present study, Field Vole populations cycled in synchrony over the whole altitude range of the study area including lowland areas with little snow cover and upland areas with prolonged snow cover.

In summary, it is difficult to find support for the proposition that yearly variations in the duration of winter snow cover are responsible for Field Vole population declines.

It is worth considering the sources of data cited by Shawyer in his analysis of field vole populations. For the most part, these were either substantially anecdotal or from isolated studies in specific areas. A comparison was made between these and a single value each year for snow duration over the whole of Britain. Although cycles often occur synchronously over large areas, many exceptions have been noted (see above). A further example is provided by this study; 1979 was clearly a year of relatively high vole numbers in the study area but was given by Shawyer as a year of low vole abundance. Extrapolation from small areas of Britain to the country as a whole cannot be made safely without validation.

A major source of data cited by Shawyer (Southern 1970), covering 21 years, did not actually refer to Field Vole but to Bank Vole *Clethrionomys glareolus*. Another, (Richards 1985) involved the results of five discontinuous studies of Field Vole spanning 30 years in small isolated patches of grassland within a large area of woodland. Richards concluded that these populations were non-cyclic and attributed this to the unusual nature of the habitats studied. Of the six remaining sources cited, five referred to a single region in south Scotland/north England. A much more rigorous approach is clearly needed if any significant understanding of Field Vole cycles in Britain is to be achieved. A major difficulty with weather hypotheses is that they are not readily amenable to experimental testing.

Barn Owl population changes in this study were not related to annual variations in the duration of winter snow cover. Shawyer's contention that winters with 20 or more days of snow cover were immediately followed by a 'crash' in Barn Owl populations is thus not supported. Indeed, the long term average snow cover for the study area was in excess of 20 days and the largest increases in the number of breeding pairs of owls followed winters of greater than 20 days snow cover.

Again the sources of data used by Shawyer are open to criticism. His analysis was based on an 'index' derived from the number of young (pulli) Barn Owls ringed each year between 1931 and 1986 as a proportion of the total number ringed of all species. It was assumed, without validation, that this index was a measure of the number of breeding pairs of owls. A major criticism of an index of this nature is that it is highly susceptible to random or deliberate changes in ringing effort, both of Barn Owls and of other species. Also, variations in productivity that occur independently of changes in population size are not taken into account. Between 1930 and 1966, the total number of young Barn Owls ringed throughout the whole of Britain averaged only around 40 a year, or the equivalent of about 12 broods. Even the smallest change in effort would have had a significant effect on an index based on these numbers. This is amply illustrated by the values given during World War II, where the index showed a twelve-fold increase over two years despite the number of Barn Owls ringed having fallen to around 15–20 a year. Shawyer's interpretation, that this represented an "overwhelming increase in the population caused by the relaxation of persecution", seems biologically improbable and the likely explanation was a change in ringers' behaviour resulting in Barn Owls being over-represented in the totals. Without corroboration, an index of this nature cannot be taken as a valid measure of owl abundance.

In this study, changes in the number of pairs of breeding Barn Owls were shown to be related to changes in the abundance of their main prey, the Field Vole. However, owl numbers decreased following the 1978/79 severe winter even though vole populations remained high, suggesting that winter weather can sometimes exert a significant influence independently of food abundance. Excluding the immediate post fledging period, most mortality of Barn Owls occurs overwinter (Glue 1973; Newton et al. 1991). Weather conditions are almost certainly involved but it is likely that the relationship is complex involving interactions between prey abundance and factors such as rainfall (amount and duration), temperature, wind and snow that affect prey behaviour and hence availability, the ability of the birds to hunt effectively and their energy requirements. The altitudinal effect following the 1978/79 winter was probably a result of the gradation in intensity of several of these factors.

The winter of 1978/79 was the most extreme since 1962/63 but, despite its severity, the number of breeding owl pairs below 150 m a.s.l was reduced by only 25% and recovery occurred within two years. An almost exactly parallel case was recorded in Utah (Marti & Wagner 1985). Following an equally extreme winter there, (1981/82) their study population fell by 43% but had recovered completely by the following year. In a third study, Muller (1989) described population changes in the Alsace-Lorraine region of France, from 1977 to 1988. Population declines associated with winter snow occurred only when snow cover exceeded 40 days. The largest decline of breeding pairs (47%) occurred following the 1981/82 winter which had 48 days snow cover but the population had recovered to 95% of the 1981 level by 1983. In all three studies, a non-breeding component of the population following the severe winters probably played an important part in the rapid recovery rates in addition to the recruitment of young birds.

The relatively small and short-lived effect of the 1978/79 winter and the low frequency with which winters of such severity occur, makes it difficult to accept the suggestion that change in winter snow duration has been an important factor responsible for long term population declines in Britain. For the idea to be tenable, much more severe winters would have to occur more frequently. From his study in Frankonia, Kaus (1977) concluded that snow cover duration was a significant factor in limiting population density only when in excess of 40 days per year, the level of the 1978/79 winter in this study.

Were snow cover the main or only factor responsible for Barn Owl declines in Britain, one would expect populations to increase during periods of mild, relatively snow-free winters. There have been three such periods during the past 40 years: 1957 to 1961,

1971 to 1976 and the late 1980s. The first coincided with a period (1956–61) when many observers noted an accelerated decline in Barn Owl numbers (Prestt 1965). Bunn et al. (1982) attempted to collate information on changes in status, including from 1971 to 1976, but their success was limited, with changes not known in 41 counties, no changes reported in six, an increase in four and a decrease in three. During this period there was exceptionally low snow cover for six consecutive winters and had snow cover been the most important factor, very substantial increases in the Barn Owl population would have been evident across the whole of Britain. That this did not occur strongly suggests that factors other than snow cover were mainly responsible for low Barn Owl numbers. There are no published data relevant to the late 1980s but any increases that may have occurred are likely to have been the result of changes in pesticide usage (see below).

The main other factors likely to be responsible for the decline in Barn Owl numbers are changes in agricultural practices and these have already been proposed as the main factors elsewhere in the Barn Owls' range (e.g. Kaus 1977; Colvin 1985). It is difficult to carry out a detailed investigation of the effects of these changes as almost all information on Barn Owl populations in Britain is anectdotal and since World War II many agricultural changes have occurred concurrently. Many changes in habitat (see O'Connor & Shrubb 1986) would have reduced the availability of suitable hunting areas and prey abundance; removal of edge habitats such as ditches and hedges, woodland clearance, the loss of overwinter stubble, stubble burning, the move from hay to silage, loss of overwinter corn stacks and increased intensity of grassland management all could have had an effect. Organochlorine pesticides almost certainly were important and there is evidence that in many areas aldrin-dieldrin poisoning played a significant part in population declines between the 1950s and late 1970s (Newton et al. 1991).

Acknowledgements

I thank Buccleuch Estates, Economic Forestry Group, the Forestry Commission and many farmers for permission to work on their land. I am grateful to Dr A.N. Charles for allowing me to use his data on vole populations up to 1979. In subsequent years, vole trapping was done by the author, Dr A. Chaudhry and I.K. Langford. Drs I. Newton, M. Marquiss and A. Village and T. Irving, J. Young, R.T. Smith and E. Fellowes provided information on nest site occupation in 1978 and 1979. S. Abbott, M. Osborne, F. Slack, P. Bell, T. Irving and

I. Langford helped with fieldwork. The research was funded by the Natural Environment Research Council, Nature Conservancy Council and World Wide Fund for Nature.

References

Batzli, G.O. 1983. Responses of arctic rodent populations to nutritional factors. *Oikos, 40*: 396–406.

Braaksma, S. 1980. Gregevens over de achteruitgang van de Kerkuil (*Tyto alba guttata*-Brehm) in west-Europa. *Wielwaal, 46*: 421–428.

Bunn, D.S., Warburton, A.B. & Wilson, R.D.S. 1982. *The Barn Owl*. Poyser, Calton.

Chitty, D. 1952. Mortality among voles (*Microtus agrestis*) at Lake Vyrnwy Montgomeryshire in 1936–9. *Philosophical Transactions of the Royal Society of London B. 236*: 505–552.

Chitty, D. & Chitty, H. 1962. Population trends among voles at Lake Vyrnwy. *Symposium Theriologica, Brno (1960)*: 67–76.

Colvin, B.A. 1985. Common Barn Owl population decline in Ohio and the relationship to agricultural trends. *Journal of Field Ornithology 56*: 224–235.

De Bruijn, O. 1979. Veodseloecologie van de Kerkuil *Tyto alba* in Nederland. *Limosa, 52*: 91–154.

Erlinge, S., Goransson, G., Hansson, L., Hugstedt, G., Liberg, O., Nilsson,T., Schantz, T. von, & Sylven, M. 1983. Predation as a factor in small rodent population dynamics in southern Sweden. *Oikos, 40*: 36–52.

Glue, D.E. 1973. Seasonal mortality in four small birds of prey. *Ornis Scandinavica, 4*: 97–102.

Hansson, L. 1979. Food as a limiting factor for small rodent numbers: tests of two hypotheses. *Oecologia, 37*: 297–314.

Hansson, L. 1984. Predation as a factor causing extended low densities in microtine cycles. *Oikos, 43*: 255–256.

Hansson, L. 1987. An interpretation of rodent dynamics as due to trophic interactions. *Oikos, 50*: 308–318.

Haukioja, E., Kapiainen, K., Niemela, P. & Toumi, J. 1983. Plant availability hypothesis and other explanations of herbivore cycles: complementary or exclusive alternatives. *Oikos, 40*: 419–432.

Henttonen, H. 1985. Predation causing extended low densities in microtine cycles: further evidence from shrew dynamics. *Oikos, 45*: 156–158.

Hestbeck, J.B. 1982. Population regulations of cyclic mammals: the social fence hypothesis. *Oikos, 39*: 157–163.

Hornfeldt, B., Lofgren, O. & Carlsson, B.-G. 1986. Cycles in voles and small game in relation to variations in plant production indices in Northern Sweden. *Oecologia, 68*: 496–502.

Jarvinen, A. 1987. Microtine cycles and plant production: what is cause and effect? *Oikos, 49*: 352–357.

Kaus, D. 1977. Zur Populationsdynamik, Okologie, und Brutbiologie der Schleiereule *Tyto alba* in Franken. *Anz. Orn. Ges. Bayern 16*: 18–44.

Krebs, C.J. 1985. Do changes in spacing behaviour drive population cycles in small mammals. *In: Behavioral Ecology. Symposium of the British Ecological Society*. Ed. by R.M. Sibley & R.H. Smith, 295–312. Blackwell, Oxford.

Krebs, C.J. & Myers, J.H. 1974. Population cycles in small mammals. *Advances in Ecological Research, 8*: 267–399.

Laine, K. & Henttonen, H. 1983. The role of plant production in microtine cycles. *Oikos, 40*: 407–418.

Lindstrom, E., Anglestrom, P., Widen, P. & Andren, H. 1987. Do predators synchronise vole and grouse fluctuations? *Oikos, 48*: 121–124.

Maercks, H. 1954. Uber den Einfluss der Witterung auf den Massenwechsel der Feldmaus (*Microtus arvalis* Pallas) in der Wesermarsch. *Deutch Pflanzenschutzdienst (Braunschweig), 6*: 101–108.

Marti, D.C. & Wagner, P.W. 1985. Winter mortality in common Barn Owls and its effect on population density and reproduction. *Condor, 87*: 111–115.

Middleton, A.D. 1930. Cycles in the numbers of British voles (*Microtus*). *Journal of Ecology, 18*: 156–165.

Middleton, A.D. 1931. A further contribution to the study of British voles (*Microtus*). *Journal of Ecology, 19*: 190–199.

Mihok, S., Turner, B.N. & Iversonj, S.L. 1985. The characteristics of vole population dynamics. *Ecological Monographs, 55*: 399–420.

Muller, Y. 1989. Fluctuations d'abondance de la Chouette effraie (*Tyto alba*) en Alsace-Lorraine (France) de 1977 a 1988. *Aves, 26*: 133–141.

Myllimaki, A., Christiansen, E. & Hansson, L. 1977. Five year surveillance of small mammal abundance in Scandinavia. *EPPO Bulletin, 7*: 385–396.

Newton, I., Wyllie, I. & Asher, A. 1991. Mortality causes in British Barn Owls *Tyto alba*, with a discussion of aldrin-dieldrin poisoning. *Ibis, 133*: 162–169.

O'Connor, R.J. & Shrubb, M. 1986. *Farming and Birds*. Cambridge, Cambridge University Press.

Pearson, O.P. 1966. The prey of carnivores during one cycle of mouse abundance. *Journal of Animal Ecology, 35*: 217–233.

Prestt, I. 1965. An enquiry into the recent breeding status of the smaller birds of prey and crows in Britain. *Bird Study, 12*: 196–220.

Richards, C.G.J. 1985. The population dynamics of *Microtus agrestis* in Wytham 1949–1978. *Acta Zoologica Fennica. 173*: 35–38.

Schindler, U. 1972. Massenwechsel der ermause, *Microtus agrestis* in Sud-Neidersachen von 1952–1971. *Zeitschrift fur Angewandte Zoologie, 59*: 189–205.

Shawyer, C.R. 1987. *The Barn Owl in the British Isles*. London, The Hawk Trust.

Schonfeld, M., Gibrig, G. & Strum, H. 1977. Brietrage zur Populationsdynamik der Schleiereule *Tyto alba*. *Hercynia, 14*: 303–351.

Southern, H.N. 1970. The natural control of a population of Tawny Owls *Strix aluco*. *Journal of Zoology London, 162*: 197–285.

Taitt, M.J. & Krebs, C.J. 1983. Predation, cover and food manipulations during a spring decline of *Microtus townsendii*. *Journal of Animal Ecology, 52*: 837–848.

Taitt, M.J. & Krebs, C.J. 1981. The effect of extra food on small rodent populations: 2 Voles (*Microtus townsendii*). *Journal of Animal Ecology, 50*: 125–137.

Tapper, S. 1979. The effect of fluctuating vole numbers (*Microtus agrestis*) on a population of weasels (*Mustela nivalis*) on farmland. *Journal of Animal Ecology, 48*: 603–617.

Tast, J. & Kalela, O. 1971. Comparisons between rodent cycles and plant production in Finnish Lapland. *Ann. Acad. Sci. Fenn. A, IV Biologica 186*: 1–14.

Taylor, I.R., Dowell, A. and Shaw, G. 1992. The population ecology and conservation of Barn Owls *Tyto alba* in coniferous plantations. *In*: C. Galbraith, S. Percival & I.R. Taylor (eds.) *The ecology and conservation of European owls*. UK Nature Conservation No. 5: 16–21. Peterborough, Joint Nature Conservation Committee.

Methods of studying the long-term dynamics of owl populations in Britain

S. Percival

Percival, S. 1992. Methods of studying the long-term dynamics of owl populations in Britain. In: *The ecology and conservation of European owls*, ed. by C.A. Galbraith, I.R.Taylor and S. Percival, 39-48. Peterborough, Joint Nature Conservation Committee. (UK Nature Conservation, No. 5.)

This paper reports on the information that can be extracted from the BTO's Nest Records and Ringing Scheme data banks to elucidate population trends in Barn Owls and Tawny Owls. Using appropriate analyses, it was possible to extract estimates of breeding success and survival rates to examine changes both in time and between different regions of the country. Some recommendations are made about how the Nest Records scheme might be improved to increase further the value of the data collected by the BTO membership, using the detailed recording techniques developed during the BTO Owls Project.

Preliminary results suggest that the Barn Owl population in Britain may be in the early stages of recovery. In the last 12 years both survival rates and breeding success appear to have been increasing. Tawny Owl numbers, on the other hand, appear to have remained fairly stable in comparison, though there may have been a slight increase in recent years.

Steve Percival, British Trust for Ornithology, The Nunnery, Nunnery Place, Thetford IP24 2PU

Present address: The Northumbrian Water Ecology Centre, The Science Complex, Sunderland University, Sunderland SR1 3SD

Introduction and aims

The British Trust for Ornithology (BTO) Owls Project was set up in October 1987 as a two and a half year study to investigate the population dynamics of Barn Owls *Tyto alba* and Tawny Owls *Strix aluco* in Britain. It has been claimed that Barn Owl numbers have declined greatly in the last 50 years (Shawyer 1987), though there is no satisfactory quantitative information available to identify how large this decline has been and whether it is still continuing. Over the same period the Tawny Owl has been reported to have maintained steady population levels (Marchant *et al.* 1990). The project was established through the sponsorship of four chemical companies, Ciba-Geigy, ICI, Shell and Sorex, to look at the databanks held at the BTO to obtain more information about these two owl species' population trends. The specific aims were:

a. to examine trends in owl breeding performance, survival and dispersal;

b. to investigate changes in the birds' population patterns in relation to weather, land use, food supply and pesticide use;

c. to establish a baseline of owl abundance and population performance for the future monitoring of owl populations.

The BTO holds several long-term databases, collected by its membership, which can provide information about owl population dynamics. The Nest Record Scheme involves the systematic collection of data from nest visits through the breeding season and hence can be used to measure breeding productivity. Data from the Ringing Scheme give information about mortality and dispersal which, when integrated with the breeding data, can enable population processes to be investigated in detail.

The purpose of this paper is to outline the data gathering and analytical methods used in the project. It looks both at the historic data already held at the BTO, and the development of additional methods for collecting more detailed data on certain aspects of owl biology. Some preliminary results are presented on some of the overall trends in population patterns.

Breeding performance: The BTO Nest Record Scheme

The Nest Record Scheme has been collating data on bird breeding performance for the last 50 years. Though sample sizes are small in the early years, it is possible to analyse the data to look at long-term changes in breeding performance for both the Barn Owl and the Tawny Owl. Sample sizes are larger for the Tawny Owl than the Barn Owl in most years reflecting the former species' greater abundance. The annual totals of nest record cards submitted for each species are given in Figure 1.

39

Barn Owl

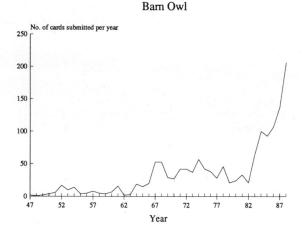

Tawny Owl

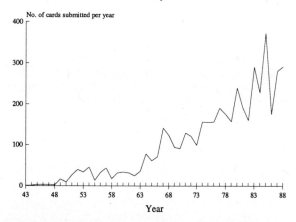

Figure 1. The total numbers of nest records cards submitted annually for (a) the Barn Owl and (b) the Tawny Owl, 1944–88.

There is a wide geographical spread in the data, enabling analysis to be made of breeding performance on a spatial basis. Figure 2 shows the total number of nest records submitted in each of the Meteorological Office regions (ones of similar climatic conditions) of Britain.

It is important that the results from analyses of these extensive data sets are treated with caution, as the information has been collected from a wide variety of habitats and nest sites by many different observers. Sampling strategies have not been rigourously assigned and the data could therefore be subject to bias. For example many records come from nestbox schemes which may often be sited in prime habitat to encourage as many owls as possible to use them, rather than in a random sample covering all the habitats that the birds use. Such potential bias should always be considered as a possible explanation for the trends observed and the results should not be treated as absolute measures of owl breeding success. As long as any biases are not unevenly distributed it should still be possible to compare between groups to examine spatial and temporal patterns.

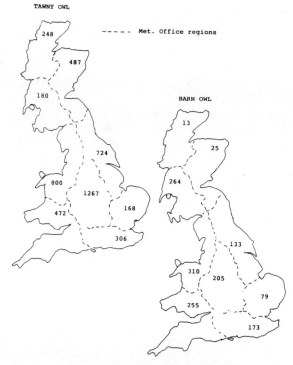

Figure 2. The total numbers of nest record cards submitted in each Meteorological Office region of Britain for Tawny Owl and Barn Owl, 1944–88.

Historically, the Nest Record Scheme has concentrated on simply recording the contents of the nest at each visit through the breeding cycle. No encouragement was made to observers to determine the age of the nest accurately, nor to plan their nest visits to maximise the usefulness of the data obtained on each visit. The only information on breeding performance available to the analyst from each nest is the contents at each visit and a subjectively assessed nest status code (for example a rough guess as to the age of the chicks).

The first step in the analysis was to obtain an estimate of the timing of breeding for each nest. This was necessary to preclude the possibility that genuine losses could be confused with birds which had fledged successfully. It also provided useful information for identifying incorrectly recorded information (for example, the number of eggs increasing after the end of the laying period) and addled eggs (which had not hatched by their predicted hatching date). The date of the first egg hatching was calculated at two quality levels:

1. If the nest was visited at a time when the nest could be aged accurately, for example during the hatching period, this information was used to give an "accurate estimate". Where several accurate estimates could be obtained from one nest an average value was used in the further analyses: differences in the hatching interval of

the eggs could give rise to some variation in these estimates, though such discrepancies were usually only of one or two days.

2. For nests which did not have such accurate information, the upper minimum first hatch date and the lower maximum first hatch dates were calculated for each visit to give a window between which the actual first hatch date could lie. The highest minimum and the lowest maximum estimates were then taken for each nest and averaged to give the first hatch date, providing these estimates were not more than 10 days apart. Nests which could not be aged to this degree of accuracy were classified "unaged" and discarded from further analyses.

The next step was to estimate the measures of breeding success for each nest:

a. the number of eggs laid;

b. the number of those which hatched;

c. the number of chicks which survived the first half of the fledgling period (11 and 22.5 days for the Tawny and Barn Owl respectively).

d. the number of chicks which survived to the end of the fledgling period (22 days for the Tawny Owl, 45 days for the Barn Owl). Though these periods were rather shorter than the real fledgling periods (averaging 28 and 60 days for the two species respectively), examination of the data showed them to be unreliable after these dates for the separation of real losses and apparent ones due to birds fledging successfully. Intensive studies on both Barn Owls (Taylor pers. comm.) and Tawny Owls (Petty pers. comm.) have shown there to be generally low mortality of chicks after these stages in the fledgling period. Monitoring of the survival of chicks after these dates was continued by ringing and subsequent recovery.

Each of these four measures was calculated by taking the nest contents at the visit closest to that date, corrected to the exact date by using the appropriate daily survival rate. The latter was calculated using the Mayfield method (1961, 1975), splitting the nesting period into three (egg, and the first and second halves of the fledgling period). This method assumes that survival rate is constant within the period being analysed. Splitting the data up in this way minimised the possibility that this assumption was violated. Thus if a nest was visited four days before hatching and contained three eggs, the estimate of the number of eggs hatched would be:

$$3 \times (\text{daily egg survival rate})^4$$

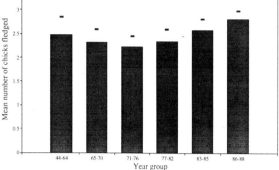

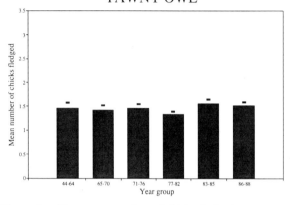

Figure 3. The mean number of chicks fledged per breeding attempt in the Barn Owl and the Tawny Owl in different year periods. Data are from the whole of Britain. 95% confidence limit bars are given on each data set.

Figure 3 shows the results of a preliminary analysis to investigate the long-term trends in breeding productivity on a national basis. The mean estimates of the number of chicks fledged per breeding attempt (the fourth of the measures discussed above) have been grouped together into means for six-year periods, to smooth out any effects due to short-term fluctuations in small mammal populations (their main food supply); the last six years have been split into two three-year periods as there are more data available then, and the data prior to 1965 have been grouped together as they are insufficient to allow any more detailed analysis. The Barn Owl fledged significantly different numbers of chicks per pair across the year groups (ANOVA, $F_{5,757} = 4.18$, $p < 0.001$). Productivity was lowest in 1971–76 but has shown a considerable increase since then. The Tawny Owl showed a similar pattern with a small but significant depression in the breeding productivity during the 1977–82 period and a subsequent increase (ANOVA, $F_{5,2558} = 4.98$, $p < 0.001$).

The overall differences in the breeding performance of owls in each of the nine Meteorological Office

Table 1. Regional variation in number of owl chicks fledged per nesting attempt, 1983–88.

	Barn Owl			Tawny Owl		
	Mean	SE	n	Mean	SE	n
NW Scotland				1.66	0.08	107
E Scotland				1.57	0.07	199
NE England	1.83	0.25	15	1.46	0.06	176
E Anglia	2.20	0.29	12	1.63	0.13	31
Midlands	2.43	0.29	25	1.48	0.05	247
SE England	3.13	0.17	53	1.45	0.08	83
SW Scotland	3.00	0.12	134	1.50	0.11	56
NW England/N Wales	2.64	0.16	70	1.67	0.10	95
SW England/S Wales	2.40	0.14	98	1.48	0.11	66

regions are summarised in Table 1. Only data from 1983–88 have been included so that the results are not confounded by temporal effects. Regional differences were tested using a one-way analysis of variance. Barn Owl breeding success was found to differ significantly between regions ($F_{6,400} = 3.23$, $p < 0.01$): birds were most successful in SE England and SW Scotland. The Tawny Owl showed no such geographical variation ($F_{8,1051} = 1.12$, $p > 0.05$, not significant).

A high proportion of the submitted records could not be used in any analysis because of their lack of accurate nest ageing information. Over 40% of the submitted nest records for both species failed to provide sufficient information for the calculation of the number of chicks fledged. A modified recording form (Figure 4) was therefore introduced in 1988, with the aim of improving the quality of the data recorded whilst visiting owl nests. This sought to encourage observers:

a. to take measurements of nest contents in addition to just recording the numbers of eggs and young, so that standard egg density and chick growth curves could be used to age the nest more accurately;

b. to make more frequent visits to nests and use the ageing information from the egg density and chick growth curves to plan their visits to maximise the usefulness of the data they collect. It was recommended that visits should be made to nests once per week through as much of the incubation and fledgling periods as possible.

In 1988 the use of this new form led to an increase in the acceptance rate of owl nest records to over 80% of those submitted.

The egg density curves for the two species derived from the measurement of eggs of known age in 1988 and 1989 are shown in Figure 5. The data have been plotted as means for each five-day period with 95% confidence limits shown. Densities were calculated using a standard formula which takes into account egg shape:

Density = Weight/(0.507 × breadth^2 × length)

The two species show a similar pattern of decline in density through incubation. More data were collected for the Tawny Owl so the confidence limits are closer to the mean. These curves can now be used to predict the stage of incubation of any egg provided its weight, length and breadth have been measured.

Figures 6 and 7 show the chick growth curves for weight, wing and head and bill length with age for the Barn Owl and the Tawny Owl respectively. As for the egg density data, means for each five-day period have been plotted with the 95% confidence limits around them. These curves can now be used to age chicks from their measurements. Confidence limits are again smaller for the Tawny Owl as more data were collected for this species.

Both wing, head and bill length were accurate predictors of chick age. Wing length was particularly useful in the latter stages of the fledgling period, when the head and bill length tended to level off in both species. Head and bill length on the other hand allowed very young chicks to be aged, when wing length was less reliable (as feathers had not yet started to grow).

Weight was not such a useful indicator of chick age: confidence limits were wider and hence variation between chicks of the same age greater. Weight will be much more dependent on food supply and therefore of more use in assessing body condition rather than ageing the chicks.

Adult and post-fledging juvenile survival rates: the ringing scheme

The BTO Ringing Scheme, like that for nest recording, has been collating data for many years. The earliest records date back to 1909. Information

BTO OWLS PROJECT

Nest Record Sheet

Please return to:- Steve Percival, BTO, Beech Grove

Tring, Herts, HP23 5NR.

Species: TAWNY	Observer: STEVE PERCIVAL	Year: 1989

Locality: ASHLEY GREEN	County: BUCKS	Grid ref: SP 834 099

Nest site: Cavity in dead birch tree	Altitude: 205 m.	Egg measurements:-

Height above ground: 5 m.

Habitat:
PRIMARY 1113 SECONDARY 5112

Egg no.	Length (mm)	Breadth (mm)
1	46.9	38.2
2	47.1	38.0
3	46.5	38.3
4	46.1	37.7

Adult details: ♀ caught 1/5.
CONTROL - GK90805 - ringed at same site as breeding adult, 26.4.88.
WEIGHT 469 g. WING 270 mm. HEAD+BILL 70.5 mm

Date	Time	Eggs	Young	Measurements & comments
10.3	18.00	1	0	Egg wt = 37.0 g. ♀ flushed.
15.3	17.10	4	0	Egg wts : (1) 34.6 (2) 37.6 (3) 37.0 (4) 35.2 ♀ flushed.
9.4	13.30	3	0	Egg no. 4 missing. ♀ flushed.
17.4	13.00	0	2	WEIGHT WING H+BILL / Chick (1) 139 44 48.5 / " (2) 130 34 48.5 ♀ flushed.
24.4	15.00	0	2	Chick (1) 243 89 60.0 / " (2) 243 71 56.0
1.5	17.30	0	2	" (1) 286 135 61.5 RINGED GH46776 / " (2) 295 119 61.0 —77
8.5	18.10	0	2	" (1) 272 148 63.5 / " (2) 312 136 62.5
15.5	17.30	0	0	Young heard calling in copse nearby.

OUTCOME OF NEST

If you have positive evidence that the young left safely, or of failure, please put a cross (☒) in the appropriate box or boxes below. Otherwise mark the appropriate "Outcome Unknown" box.

OUTCOME UNKNOWN { Because evidence for or against success is inconclusive ☐ Because observations on nest were not continued ☐

EVIDENCE FOR SUCCESS
- Young capable of leaving nest when last seen ☐
- Young: seen leaving naturally ☐ left when approached ☐
- Young seen and/or heard near nest ☑
- Parent bird(s): giving alarm calls ☐ carrying food ☐
- in nest: hatched shells ☐ feather scale ☐ droppings ☐
- Any other evidence:-

EVIDENCE FOR FAILURE
{ Nest: empty ☐ damaged ☐ fallen ☐ flooded ☐ removed ☐
Eggs: damaged ☐ deserted ☐ all infertile or addled ☐
Young: all dead, uninjured ☐ all dead, 1 or more injured ☐
Any other evidence (e.g. type of weather causing failure, species of predator if seen, etc.):-

Figure 4. The BTO Owls Project nest recording form, filled out to illustrate the "ideal" record.

Barn Owl

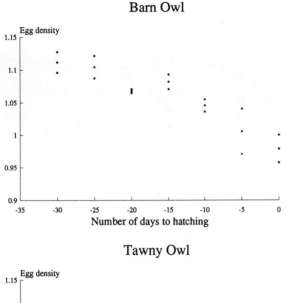

Tawny Owl

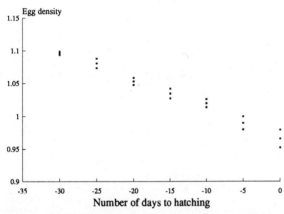

Figure 5. The relationship between egg density and hatching date for (a) the Barn Owl and (b) the Tawny Owl, showing how egg measurements can be used to estimate hatching date (where egg density = weight/0.507 × (breadth)2 × length). The data are 5-day means with the 95% confidence limits shown.

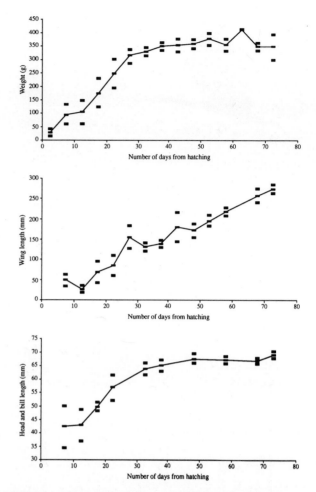

Figure 6. The relationship between chick biometrics and time since hatching for the Barn Owl, showing how chick measurements can be used to estimate hatching date. The data are plotted as five day means and 95% confidence limits are shown.

can be extracted about survival and dispersal over a long-term period. The total number of ringed birds that have been recovered for each species each year since 1930 is shown in Figure 8. Although approximately equal numbers have been recovered for the two species, more Tawny Owls have been ringed. The percentage of ringed birds recovered is much lower for the Tawny Owl (7.8%) than for the Barn Owl (16.0%), according to totals published by Mead & Clark (1988).

The regional spread of the ringing data is also wide. Figure 9 shows the numbers of the recoveries of each species in each of the Meteorological Office regions. The data are not sufficient, however, to carry out a regional analysis of survival rates on that scale, so for this initial preliminary analysis the results are presented on a national basis. The main problem with the sample sizes is the relatively small numbers of adults (necessary to provide a robust estimate of

adult survival) which have been ringed: a large proportion of owls are ringed as chicks in the nest (for example 92% of Barn Owls and 86% of Tawny Owls ringed in 1987 were chicks, Mead & Clark 1988).

Traditional methods of ring recovery analysis to determine survival rates (Haldane 1955) are inappropriate for owls, as they assume that survival rate is constant. They cannot allow for the short-term variation in survival rates which has been shown to occur in many owl species (for example in Barn Owls by Sauter 1956). A more suitable approach should allow survival rates to be both age- and time-specific. Recent advances in analytical technique and computer software now make such an approach possible. The SURVIV package (White 1983) was used for the calculation of survival rates in this case. The data were analysed to find the simplest model (and therefore the one which would give greatest precision in the estimates that it produced) which still

44

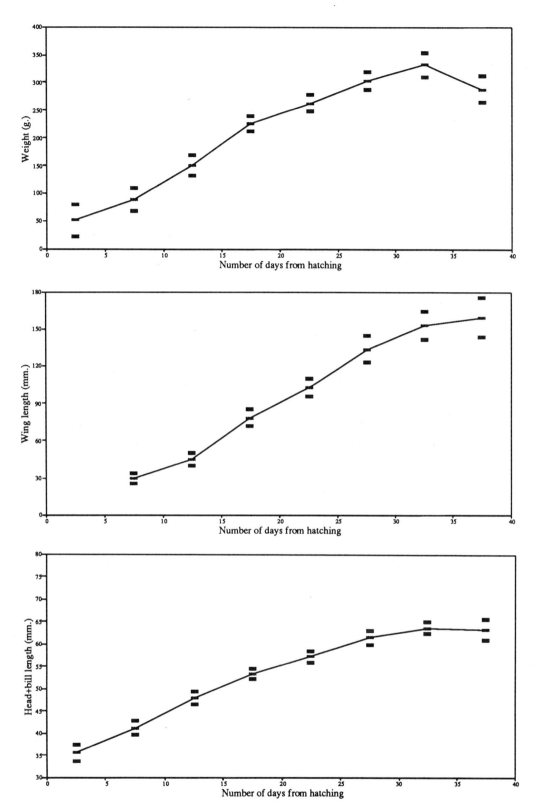

Figure 7. The relationship between chick biometrics and time since hatching for the Tawny Owl, showing how chick measurements can be used to estimate hatching date. The data are plotted as five day means and 95% confidence limits are shown.

gave a significant fit to the data (with a chi-squared goodness-of-fit test, using $p < 0.05$ as the threshold level of significance). Firstly the analysis was carried out using a general model where all adult and first-year survival and recovery rates were estimated separately, then the model was progressively refined so that more parameters (annual recovery and survival rates) were set equally. The final model

45

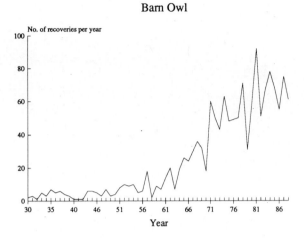

Barn Owl

No. of recoveries per year

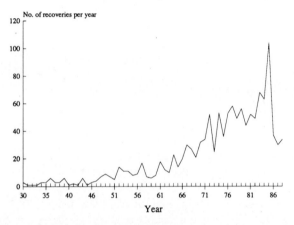

Tawny Owl

No. of recoveries per year

Figure 8. The number of ringed birds recovered annually from 1930 to 1988 for (a) the Barn Owl, (b) the Tawny Owl.

choice, for both species, was one which comprised:

Year-specific first-year survival rate
Year-specific adult survival rate
Constant adult recovery rate
Constant first-year recovery rate

Preliminary results of this analysis on a national scale are presented in this paper to illustrate the overall trend in owl survival rates over the last 12 years (the time period for which there are adequate data for the analysis). These results are shown in Figure 10. Survival rates for adult Tawny (r=0.455, p=0.0198) and Barn Owls (r=0.788, p=0.0002) both showed significant increases through the 12-year period, but no such significant trends were found for first year survival rates in either species (r=0.182, p=0.205 for the Tawny Owl and r=0.0121, p=0.292 for the Barn Owl). All of these survival rate trend correlation statistics are given as Kendall's tau. The sample size is 12 years in all cases.

Survival rates are not the only information that can be obtained from the Ringing Scheme data. Details

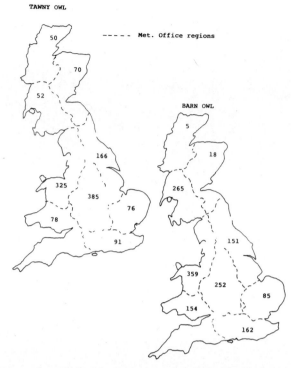

Figure 9. The total numbers of ringed birds recovered in each Meteorological Office region of Britain for Tawny Owl and Barn Owl, 1930–88.

of all birds ringed are submitted to the BTO, including the size of the brood (when the birds were ringed as chicks). The latter provides supplementary information to the Nest Record Scheme on breeding performance, which could be particularly useful where sample sizes are otherwise too small. These records of the brood size at ringing were standardised with the estimate of the number of chicks fledging per nest (from the Nest Record data) by using two statistics extracted from the Nest Record data:

a. the mean age at which chicks were ringed. The date of ringing was often recorded on the Nest Record cards so the age of those chicks could be calculated;

b. the appropriate daily chick survival rate, calculated as explained above.

The standard estimate of the number of chicks fledged could then be calculated from the recorded brood size at ringing using the following equation:

"Fledging success" = Brood size at ringing $\times$ (daily chick survival rate)^x

where x = no. of days between the end of the "fledgling" period and the mean age of the chicks at ringing;

which = 40–33 = 7 for the Barn Owl

and = 22–21 = 1 for the Tawny Owl

46

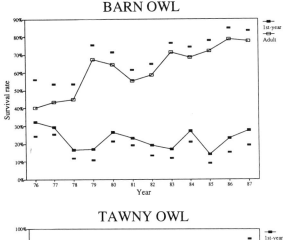

BARN OWL

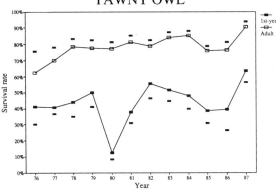

TAWNY OWL

Figure 10. Annual adult and first-year survival rates for (a) the Barn Owl and (b) the Tawny Owl during the period 1976–1987. Data analysed using the "SURVIV" package (White 1983) and assuming constant adult and first-year recovery rates. Standard error bars are plotted.

Some particular care is necessary in the interpretation of these data as only successful nests will be recorded (since failed nests will not produce any chicks to be ringed) and so will overestimate the number of chicks fledged.

Integrated monitoring: bringing the data together

By bringing together the results from the breeding performance and survival rate analyses, it is possible to gain an overall picture of owl population dynamics and to examine losses through the breeding cycle and both adult and post-fledging juvenile mortality. The technique of key-factor analysis (Varley & Gradwell 1960) serves to illustrate the importance of losses at each stage and identify those which may be affecting total population numbers. The killing power (k_n) of each stage is calculated as:

log10 (population at the start of that stage)

− log10 (population at end of stage)

The analysis started with the maximum clutch, taken as nine for the Barn Owl and five in the Tawny Owl

(Cramp 1985). The first k-factor, k_1, represented the loss between this and the actual clutch laid, caused by females failing to lay a full clutch. Subsequent losses included in the analysis were:

k_2, the mean number of eggs laid − mean number that hatch

k_3, the mean number of eggs that hatch − mean number of chicks that survive to half way through the fledging period.

k_4, the mean number of chicks that survive to half way through the fledgling period − mean number of chicks that fledge

k_5, the mean number of chicks fledged − mean number that survive to the end of their first year

k_6, the mean loss of adults from the population through the year.

k_1 through to k_4 were estimated using data extracted from the nest record cards, k_5 and k_6 from the survival rates calculated from the ringing recovery data. Figure 11 illustrates the timing of these losses through the birds' annual cycle. The total loss (K) was calculated as the sum of all $k_1 \ldots k_n$. The contribution of each stage to the overall losses was assessed by correlating each k_n with K across years: the larger the correlation coefficient the more important that stage is in determining overall losses. Figure 12 shows the results of such an analysis at the national level for the period 1976–1987 for the Barn

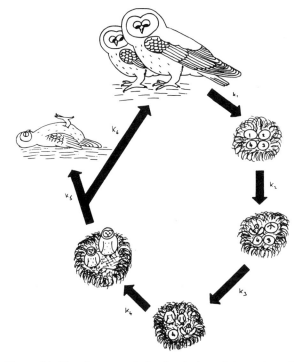

Figure 11. Key factor analysis: the timing of the action of the various k-factors through the annual cycle.

BARN OWL KEY FACTOR ANALYSIS

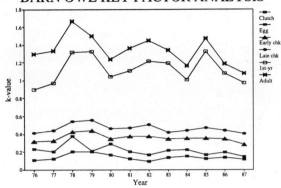

TAWNY OWL KEY FACTOR ANALYSIS

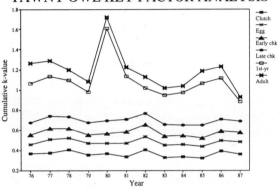

Figure 12. Key factor analyses of the Barn and the Tawny Owl in Britain, 1976–87.

Table 2. Correlation of owl population measures ($k_1...k_6$) with overall loss (K).

k-factor	Barn	Tawny
k_1 Failure to lay full clutch	0.182	0.212
k_2 Incubation losses	0.030	−0.212
k_3 Early chick losses	0.273	0.091
k_4 Late chick losses	0.091	*0.485
k_5 Post-fledging juvenile mortality	**0.578	**0.606
k_6 Adult mortality	0.212	**0.576

(All correlation coefficients are Kendall's tau. $* = p < 0.05$, $** = p < 0.01$).

and Tawny Owl respectively, and Table 2 the corresponding correlation coefficients. Table 2 shows that for both species post-fledging mortality is the most important factor affecting population levels. Adult mortality and late chick losses were also important for the Tawny Owl.

Thus the BTO long-term databanks can be used to elucidate trends in Barn and Tawny Owl population patterns. The preliminary analyses presented suggest that, contrary to some current opinion (for example Shawyer 1987), Barn Owl adult survival rate and breeding performance have both increased over the last 12 years. At the same time the results concur with current thinking on the population trend of the

Tawny Owl, one of continuing stability (Marchant *et al.* 1990). Using the techniques outlined in this paper it has been possible to extract information from the BTO long-term databases which can illustrate overall population patterns. These form the basis for further analyses with other data, for example on the bird's environment, enabling interpretation of the patterns.

Acknowledgements

The data on which this paper is based has been collected by many contributors to the BTO's Nest Records and Ringing Schemes. I thank everyone who has sent in owl data and encourage them to continue in their valuable work.

Many people have given me help and encouragement with this work: I would particularly like to thank Colin Galbraith, Tony Hardy, Ian Langford, Steve Petty and Iain Taylor. Rob Fuller provided many useful comments on a draft of the paper. Thanks also for comments from an anonymous referee. Secretarial support was provided by Tracey Percival, Liz Murray and Sue Taylor, and Nichola Bayman had the unenviable task of inputting 25,000 ringing records onto the computer. Also a big thank-you to Philip Burton for allowing me to use his nestboxes for detailed nest monitoring, and to Patrick Thompson and Tracey Percival for logistical support carrying ladders around Hertfordshire and Buckinghamshire.

The project was sponsored by Ciba-Geigy, ICI, Shell and Sorex. Long-term support to the Nest Records and Ringing databases has been given by the Nature Conservancy Council.

References

Cramp, S. 1985. *Birds of Europe, the Middle East and North Africa.* Volume IV. Oxford University Press, Oxford.

Haldane, J.B.S. 1955. The calculation of mortality rates from ringing data. *Proceedings of the 11th Ornithological Congress, Basel.* 454–458.

Marchant, J., Hudson, R.H., Carter, S. & Whittington, P.A. 1990. *Population trends in British breeding birds.* BTO, Tring.

Mayfield, H. 1961. Nesting success calculated from exposure. *Wilson Bulletin, 73:* 55–261.

Mayfield, H. 1975. Suggestions for calculating nest success. *Wilson Bulletin, 87:* 456–466.

Mead, C.J. & Clark, J.A. 1988. Report on bird ringing in Britain and Ireland for 1987. *Ringing and Migration, 9:* 169–204.

Sauter, U. 1956. Barn Owl movements and survival in relation to weather and rodent abundance. *Vogelwarte, 18:* 109–151.

Shawyer, C.R. 1987. *The Barn Owl in the British Isles: its past, present and future.* The Hawk Trust, London.

Varley, G.C. & Gradwell, G.R. 1960. Key factors in population studies. *Journal of Animal Ecology, 29:* 399–401.

White, G.W. 1983. Numerical estimation of survival rates from band-recovery and biotelemetry data. *Journal of Wildlife Management, 47:* 716–728.

Effects of new rodenticides on owls

I. Newton & I. Wyllie

Newton, I., & Wyllie, I. 1992. Effects of new rodenticides on owls. *In: The ecology and conservation of European owls*, ed. by C.A. Galbraith, I.R.Taylor and S. Percival, 49-54. Peterborough, Joint Nature Conservation Committee. (UK Nature Conservation, No. 5.)

Rats and mice in many regions have become resistant to warfarin and other contemporary anticoagulant poisons. This has led to the development of other anticoagulant chemicals for rodent control, the so-called "second generation" rodenticides, namely difenacoum, bromadiolone, brodifacoum and flocoumafen. These new compounds are much more toxic than warfarin, and some are also more persistent, so they have greater capacity to cause "secondary poisoning" in rodent-predators.

Because Barn Owls *Tyto alba* frequently roost and nest in farmsteads and other buildings, and feed on commensal rats and mice caught in and around the buildings, they might be especially vulnerable from secondary poisoning following rodenticide use. Apart from occasional reports of Barn Owls found dead near buildings where new rodenticides were used (Shawyer 1987), until recently little information was available from Britain on which to make any assessment of the risk to these owls posed by rodenticide use. However, in Malaya, the replacement of warfarin by coumachlor and brodifacoum to control rats in oil palm plantations is said to have caused massive decline of Barn Owl populations through secondary poisoning, with dead and dying owls found with appropriate symptoms (Duckett 1984); in the United States field trials with brodifacoum resulted in the deaths of Screech Owls *Otus asio* (Hegdal & Colvin 1988); and in Switzerland the use of bromadiolone in one area to control Ground Voles *Arvicola terrestris* caused a mass mortality incident involving an estimated 185 Buzzards *Buteo buteo*, 25 Red Kites *Milvus milvus*, one Goshawk *Accipiter gentilis* and several predatory mammals (Pedroli 1983; Beguin 1983). In this paper, we review the main second generation rodenticides now in common use in Britain and elsewhere, describe the rodenticide contents of Barn Owls found dead in Britain, and present the results of some trials undertaken to determine the toxicity of two rodenticides to captive Barn Owls. We also discuss the findings from other studies on various owl species. Ultimately, this information should help towards assessing the likely impact of rodenticides on wild owl populations.

I. Newton & I. Wyllie, Institute of Terrestrial Ecology, Monks Wood Experimental Station, Abbots Ripton, Huntingdon, Cambridgeshire PE17 2LS

The chemicals

Warfarin, the first anticoagulent rodenticide, was developed in 1942 from the naturally occurring compound coumarin, first isolated from clover. It was introduced for use in Britain from 1952, and was soon followed by other similar chemicals. It is still the most commonly used rodent poison, but two other "first generation" rodenticides are still available here, namely coumatetralyl (trade name: Racumin) and chlorophacinone (Drat). With all these chemicals, several feeds are needed, and death follows some days later. So far as is known, these chemicals present no serious threat to rodent predators (see Townsend *et al.* 1981 for work on Tawny Owls *Strix aluco*).

Rats first began to show genetic resistance to warfarin in one area in the late 1950s. Resistance then developed independently in rats and mice in other areas and spread throughout the 1960s. This stimulated the development of new, more potent, anticoagulants, beginning with difenacoum (on sale in Britain from 1975), followed by bromadiolone (from 1980), brodifacoum (from 1982) and

flocoumafen (from 1986). These chemicals are also coumarin-based. Difenacoum is about 100 times more toxic than warfarin, and brodifacoum about 600 times more. Thus not only will these chemicals kill rodents resistant to warfarin they will do so with a single feed, allowing "pulsed" baiting as opposed to the sustained baiting required with warfarin. They have therefore found wide acceptance. The toxicity values quoted for these compounds in Table 1 represent the amount of active ingredient, given as a single oral dose, which was needed to kill 50% (LD_{50}) of a sample of captive animals. The dose is expressed as mg per kilogram of animal body weight.

All these chemicals are more toxic to rodents than to the other mammals and bird species that have been tested (Worthing & Walker 1987). In general, larger species can withstand much larger doses, relative to their body weight, than can smaller species. Brodifacoum is the most toxic of the four.

Another important aspect of these chemicals, which influences their effects on rodent-predators, is their persistence in the bodies of animals which are

Table 1. Relative toxicity of four second generation anticoagulants.

| Chemical | Trade names | Acute lethal dose (LD_{50}, in mg/kg body weight) | | | | | |
		Mouse	Rat	Rabbit	Cat	Chicken	Mallard
Difenacoum	Neosorex, Ratak, Ratrick, Fentrex	0.80	1.80	–	100	–	–
Bromadiolone	Maki, Bromard, Bromatrol, Bromex, Deadline, Slaymore	1.75	1.13	1.0	–	–	–
Brodifacoum	Sorex brodifacoum, Talon B, Ratax plus Klerat, Volid	0.40	0.27	0.3	ca. 25	4.5	2.0
Flocoumafen	Storm, Strategem	0.8	0.25	0.7	–	>100	ca. 100

Difenacoum was introduced by Sorex and later ICI, bromadiolone was made by Lipha SA, brodifacoum was developed by ICI and introduced by Sorex, and flocoumafen was introduced by Shell International Chemical Company. Details from Worthing & Walker (1987).
The quoted LD_{50} values may be compared with estimated values for warfarin of 186 mg/kg for rat and 374 mg/kg for mouse, ca. 800 mg/kg for rabbit and >1000 mg/kg for chicken (Thomson 1976, Hagen & Radomski 1953). Alternatively, rats are killed by five daily warfarin doses of 1 mg/kg, cats by five doses of 3 mg/kg, and pigs by five doses of 1 mg/kg; chickens are more resistant (Worthing & Walker 1987).

exposed but not killed. Little published information is available on this aspect, but brodifacoum is known to persist in the bodies of wild animals for up to several months (Rammell *et al.* 1984). Such animals could therefore represent a source of residues to their predators for long after their initial exposure. So far as is known, all anticoagulant rodenticides act in a similar way. The site of action is the liver, where four blood-clotting proteins are synthesised under the action of vitamin K. This vitamin is in turn continually regenerated in the liver in a cyclical process, involving three specific enzymes. The anticoagulant binds to one of these enzymes (vitamin K epoxide reductase). This prevents the synthesis of vitamin K, and the dependent blood-clotting proteins, which gradually become depleted in the blood stream. Thus death occurs several days after exposure from internal haemorrhaging. Vitamin K is an effective antidote.

Screening of Barn Owl carcasses

In order to gain some idea of the extent to which Barn Owls in Britain were exposed to the new rodenticides, advertisements were placed in bird journals requesting the carcasses of any Barn Owls found dead. All carcasses were requested irrespective of mode of death, and stored deep frozen until they could be examined. On examination a visual post-mortem was conducted which, together with information from the sender, was used to diagnose the cause of death. Special attention was paid for signs of internal haemorrhaging. From each carcass a piece of liver tissue was extracted, and analysed for two commonly-used rodenticides, difenacoum and brodifacoum. (It is hoped in future to analyse for other rodenticides, but this has not yet been

attempted). Analysis was by high pressure liquid chromatography using the methods of Hunter (1985). The lower limit of detection for both compounds was around 0.01 μg, which was equivalent to 0.005–0.01 ppm, depending on the weight of the sample.

During 1983–89 some 145 Barn Owl carcasses were received from various parts of Britain (Newton *et al.* 1990). Of these 56% were diagnosed as road traffic victims, 10% as other accident victims, 32% as starved and 2% as shot. Only one was diagnosed as an obvious rodenticide victim.

On chemical analysis, difenacoum was detected in seven birds, brodifacoum in four and both chemicals together in a further four. In other words, about 10% of all the birds examined contained at the time of their death one or other of these two chemicals. Liver residues of difenacoum were in the range 0.005–0.106 ppm, and of brodifacoum 0.019–0.513 ppm. These are fresh weight values, corrected to allow for water loss between death and analysis (Newton *et al.* 1990). The 15 contaminated birds came from England and Wales, ranging from the southeast to Humberside, and from eastern England to Wales and the Isle of Man. They were not confined to areas where rodents are warfarin-resistant. Evidently Barn Owls over much of their British range are now exposed to new rodenticides, but on their own, these findings do not indicate what proportion of contaminated birds are likely to have died of rodenticide poisoning.

Laboratory tests

In order to assess the significance of the residues found in wild owls, and whether they were likely to have been lethal, we fed captive Barn Owls on dead rodenticide-poisoned mice, and monitored the effects. The results are described in detail in Newton *et al.* (1990), and only a summary is given here.

Laboratory mice, fed for one day on difenacoum or brodifacoum bait, died 2–11 days later. Some of these dead mice were analysed to determine their rodenticide contents, and others were fed to captive Barn Owls. Six owls were fed for one day on difenacoum-killed mice (three per owl), and another six owls were fed for one day on brodifacoum-killed mice (three per owl). After dosing, blood samples were taken periodically from the owls to monitor coagulation times. This procedure served to assess effects of the rodenticides, and also to indicate the recovery times. Any owls which survived the one day feeding trial were later fed for three successive days on rodenticide-poisoned mice, and those which recovered from this treatment were then fed for six successive days on poisoned mice.

The six owls fed on difenacoum-poisoned mice all survived the 1-day, 3-day and 6-day treatments. After the 1-day treatment, all six owls were blood-sampled 5–9 days later, and coagulation times were normal. After the 3-day treatment, the blood of one bird taken three days later would not coagulate, but blood from all birds was normal in this respect 9–23 days later. Hence, with difenacoum, the effects were temporary, and did not lead to death. No external haemorrhaging was seen.

Of the six owls fed on brodifacoum, four died 6, 10, 11 and 17 days after the one day treatment. Their livers contained 0.63–1.25 ppm in fresh weight of brodifacoum. The two survivors also survived the 3-day and 6-day treatments. Some of these owls bled periodically from the mouth; blood taken from two birds would not coagulate nine days after the end of feeding, and nor would blood taken from one of these birds (a survivor) 38 days later. However, the blood from the other bird seemed normal 16 days after treatment. Hence, brodifacoum was much more toxic to Barn Owls, with longer-lasting effects, than difenacoum.

The mice that died weighed an average of 35 g each and were found on analysis to have an average of 10.2 µg difenacoum or 15.4 µg brodifacoum in their body. As the owls that died each ate three mice, they would each have consumed in the one day trial an average of 46.2 µg brodifacoum. This was equivalent to a dose of 0.15–0.18 mg/kg body weight for the birds concerned. On analysis, the livers of the dead owls contained an average of about 4.6 µg brodifacoum, about one tenth of the estimated total consumption. On post-mortem, haemorrhaging was found along both sides of the keel, and around the brain, heart, lungs or gut, and in two birds blood filled the body cavity. Three of the wild owls had brodifacoum in their liver at à level (0.30–0.52 ppm) close to that which killed our captive owls. In the five other wild owls which contained brodifacoum, the level was an order of magnitude lower (0.02–0.07 ppm).

Discussion

Problems and deficiencies in the data make it impossible at this stage to assess the risk that new rodenticides pose to Barn Owls in Britain, let alone the proportion dying, or the effect on population trend.

1. Of the four second generation rodenticides now in common use, we have so far examined only two for their occurrence in, and effects on, Barn Owls.

2. We have no information on how long after exposure these chemicals persist in the bodies of live owls, so cannot say whether the 10% contaminated owls represented single or multiple exposure over lifetimes (as one extreme) or single exposures immediately before death (as the other). If the chemicals at sub-lethal levels are short-lived in owl bodies, results imply that more than 10% of owls would be exposed during their lifetimes.

3. The sample of wild owls was probably biased against rodenticide victims. Death from anticoagulant is slow and preceded by lethargy, so affected owls are most likely to die at their roosts, in tree holes or roof spaces, where they would be unlikely to be found by people. Almost all the owls we obtained for analysis had been found in the open, mostly near roadsides.

4. Comparison of residues in wild owls with those in the poisoned captive birds would suggest that no more than three out of eight wild owls containing brodifacoum had died of brodifacoum poisoning. However, it is also possible that wild owls die at lower levels of rodenticide than were needed to kill captive birds. As wild birds are more active, haemorrhaging may be more frequent and prolonged than in the captive birds which remained inactive in their individual cages, moving only to eat or when disturbed. Hence, caution is needed in extrapolating directly from captive owls to wild ones.

There is also the possibility that rodenticides in the bodies of dead owls break down slowly over time. In this case the levels found in our long-stored carcasses (up to 4 years for wild birds) would have been lower than in the same individuals at death, and not comparable with the results from our captive birds, analysed on the day of death. The persistence of residues in live birds and in frozen carcasses is currently under investigation.

Despite these problems, the following conclusions seem justified on the data presented:

(1) exposure of Barn Owls to rodenticides in Britain is frequent and widespread;

(2) exposure is not restricted to areas where rodents are warfarin-resistant;

(3) owls obtain brodifacoum despite its supposed restriction to use inside buildings;

(4) poisoned rodents remain alive and available to owls for several days after exposure;

(5) three mice (possibly fewer) are enough to provide a lethal dose of brodifacoum; and

(6) brodifacoum is more toxic to Barn Owls than is difenacoum.

Some other relevant work

With captive owls

Mendenhall & Pank (1980) compared the secondary toxicity of six rodenticides to owls. Rats were killed by feeding an anticoagulant bait for five days, and were then offered to individual owls for periods of 1, 3, 6 or 10 days. For three "first generation" rodenticides, Barn Owls fed on fumarin (2), chlorophacinone (2) or diphacinone (2) survived and showed no gross poisoning symptoms. For three second generation rodenticides, three out of six Barn Owls fed difenacoum haemorrhaged but did not die, while one out of six fed bromadiolone died (at 11 days), and five out of six fed brodifacoum died (at 8–11 days). These results were therefore consistent with ours on the differential toxicity of difenacoum and brodifacoum to Barn Owls. In another trial, Mendenhall & Pank fed diphacinone-poisoned rats to one Saw-whet Owl *Aegolius acadicus* and three Great-horned Owls *Bubo virginianus*; the Saw-whet and two of the Great-horned Owls died at 7–14 days.

In another experiment of this type, Townsend *et al.* (1981) fed captive Tawny Owls on a diet of warfarin-killed mice for a period of three months. All four birds survived, but at the end of the period their prothrombin levels in blood were lower than normal. This experiment was undertaken because warfarin is often used in British woodlands for Grey Squirrel

Sciurus carolinensis control, and the bait is also available to small rodents. The authors concluded that Tawny Owls were unlikely to obtain a lethal dose from consuming warfarin-contaminated mice in woodlands treated for squirrel control.

In a further study, six Tawny Owls were fed on difenacoum-killed mice until they died (Anon 1982). The mice were presented after having been fed on difenacoum bait for three days, by which time residues in their bodies should have reached their peak. In the owls, symptoms of poisoning which included lethargy, unwillingness to fly, unawareness of surroundings and occasional external bleeding, were seen after about seven days. All birds died some 8–41 days after the start of dosing. On post-mortem, internal bleeding was evident in the pectoral muscles, gizzard, neck and leg joints. The liver and certain other tissues contained less than 0.2 ppm difenacoum. Ejected pellets contained up to about 0.7 ppm and faeces some 10–20 ppm. The author concluded that, although "Tawny Owls may be susceptible to secondary poisoning from difenacoum-contaminated mice, it is unlikely that wild owls would be subject to the same consistent degree of exposure to the rodenticide".

With wild owls

Two large field trials, conducted in the United States, examined the effects of brodifacoum use on wild Barn Owls and Screech Owls *Otus asio*. In the Barn Owl study, 34 owls were radio-marked in an area where 0.005% brodifacoum bait (Talon) was used on 25 out of 40 farms (Hegdal & Blaskiewicz 1984). Of 18 owls on treated farms, at least 9–12 were still present 5–62 days post-treatment, and at least 8 broods fledged from farms where poisoned rodents were available for part of the nesting cycle. On nearby control farms, 16 owls were radio-tagged and found to feed on or near the treated farms. In total 8 owls were found dead, but none from brodifacoum poisoning (though detectable residue was found in one electrocuted bird). The authors concluded that "the potential for Barn Owl mortality as a result of Talon rodenticide baiting around farms appears to be low". However, the owls in this summer study had hunted mainly away from farm buildings, in fields and marshes, where Meadow Voles *Microtus pennsylvanicus* were the major prey. In another trial, 0.001% brodifacoum bait (Volid) was applied in orchards against voles *Microtus* spp. (Hegdal & Colvin 1988). Of 32 radio-tagged Screech Owls that hunted in the orchards, eleven were found dead (together with one Long-eared Owl *Asio otus*) some 5–37 days after treatment, and secondary poisoning with brodifacoum was implied in at least six. Five of these birds contained 0.5–0.8 ppm brodifacoum in their liver (levels similar to those in our captive Barn Owls), and the sixth showed extensive haemorrhaging. Later, six other radio-marked Screech Owls were collected alive for residue analysis, and four contained detectable brodifacoum residues, at 0.3–0.6 ppm in their liver. The authors concluded that their results indicated "the need for concern when anticoagulant rodenticides, in particular newer (second-generation) compounds, are used or proposed for field rodent control".

The events recorded by Duckett (1984) for Barn Owls in Malayan oil palm plantations could be regarded as another field trial, but represented an exceptional situation. The owls concerned were living and breeding only within the plantations where the rats were being controlled, and were eating almost nothing but rats. In these circumstances, with brodifacoum the main chemical used, the recorded mass mortality and population collapse in the owls was probably inevitable.

Residue levels in rodents

Our laboratory mice contained an average of 15.4 μg of brodifacoum in their bodies after death, and three such mice (with a total of 46.2 μg brodifacoum) proved lethal to four out of six Barn Owls (see above). Comparable information on body loads is available for wild rodents. Thus, at three different application rates, voles live-trapped up to 19 days after treatment contained means of 0.35 ± 0.03, 2.07 ± 0.17 and 4.07 ± 0.20 ppm brodifacoum in their bodies (Merson et al. 1984). For a 30 g vole, these concentrations are equivalent to weights of 10.5, 62.1 and 122.1 μg brodifacoum respectively per animal. On this basis a single vole at the two highest application rates would contain enough brodifacoum to kill an owl.

Some 52 Norway Rats Rattus norvegicus, picked up dead after 'pulsed' baiting, contained an average of 1.4 ppm brodifacoum in their bodies, while 26 others collected after continuous baiting contained an average of 3.2 ppm (Dubock 1984). In a 500 g rat this would represent 700 μg and 1400 μg respectively of brodifacoum. Thus, on both baiting systems, one poisoned rat could prove lethal for a Barn Owl.

As our laboratory mice took 2–11 days to die after having eaten rodenticide, they would presumably represent a source of residues to owls for a similar period. These findings are probably typical, as Bajomi (1984) recorded periods between exposure and death of 3–11 days for laboratory mice and 2–14 for laboratory rats, while Hoppe & Krambias (1984)

recorded 1–9 days for Black Rats Rattus rattus. For most of the lag period the rodents behave normally, becoming lethargic only a few hours before death.

Discussion

Clearly, second generation rodenticides are likely to have more impact on owl populations if they are used in the wider countryside than if they are restricted to use around buildings. They are also likely to have more impact if they are used on a major, rather than a minor, prey species. In either situation, a single poisoned animal could provide a lethal dose of brodifacoum. Whether such chemicals affect Barn Owls using buildings in Britain will depend largely on where the owls are hunting. For most of the time, Barn Owls hunt mainly away from farm buildings, but commensal rats and mice are occasionally taken, and are likely to increase in the diet when voles are scarce (at low points in the vole cycle) or unavailable (during periods of snow cover). It is under these circumstances, when Barn Owls are at greatest risk from starvation, that they are likely to be at greatest risk from rodenticides too. This double risk will make the impact of rodenticides on Barn Owl populations even harder to evaluate.

Summary

1. Four second-generation rodenticides are now in common use in Britain. Like warfarin, these new chemicals are coumarin-derivatives, and act as anticoagulants. They are more toxic than warfarin, and more likely to cause secondary poisoning of rodent-predators. The most toxic is brodifacoum, at present restricted in Britain to use inside buildings.

2. During 1983–89, 145 Barn Owls found dead in various parts of Britain were analysed for residues of difenacoum and brodifacoum. Some 10% of these owls were found to contain one or both chemicals. Difenacoum was found at 0.005–0.106 ppm in liver (wet weight) and brodifacoum at 0.019–0.513 ppm.

3. Of six owls fed for one day on brodifacoum-killed mice, four died 6–17 days later. Brodifacoum was found in their livers at levels of 0.63–1.25 ppm wet weight. The two surviving owls also survived three and six day treatments. Of six other owls fed on difenacoum-killed mice, all survived one, three and six-day treatments.

4. Some other work is reviewed. In a field trial in New Jersey, in which brodifacoum baits were

used on farmsteads with nesting Barn Owls, no owls were found dead from brodifacoum poisoning. However, the owls at the time had fed mainly in fields and marshes away from farmsteads. In another trial in Virginia, in which brodifacoum baits were used in orchards, 11 out of 50 radio-tagged Screech Owls were found dead 5–37 days later, together with one Long-eared Owl. Analyses implied brodifacoum poisoning of at least six.

References

Anon. 1982. Secondary toxicity hazard to owls from difenacoum. *Pesticide Science 1981*. Ministry of Agriculture, Fisheries and Food, reference book 252(81). London, HMSO.

Bajomi, D. 1984. Hungarian experiences with "Talon-B" containing brodifacoum. *In: The organisation and practice of vertebrate pest control*, ed. by A.C. Dubock, 163–179. Haslemere, England: ICI.

Beguin, J. 1983. *Report on the chemical control of voles*. Neuchatel: The Department of Agriculture of the Republic and Canton of Neuchatel.

Dubock, A.C. 1984. Pulsed baiting - a new technique for high potency, slow acting rodenticides. *In: The organisation and practice of vertebrate pest-control*. ed. by A.C. Dubock, 105–142. Haslemere, England: ICI.

Duckett, J.E. 1984. Barn Owls (*Tyto alba*) and the "second generation" rat-baits utilised in oil palm plantations in Peninsular Malaysia. *Planter, Kuala Lumpur, 60*: 3–11.

Hagen, E.C. & Radomski, J.L. 1953. The toxicity of 3-(acetonylbenzyl)-4-hydroxycoumarin (Warfarin) to laboratory animals. *Journal of the American Pharmaceutical Association, 52*: 379–82.

Hegdal, P.L. & Blaskiewicz, R.W. 1984. Evaluation of the potential hazard to Barn Owls of talon (brodifacoum bait) used to control rats and house mice. *Environmental Toxicology & Chemistry 3*: 167–179.

Hegdal, P.L. & Colvin, B.A. 1988. Potential hazard to Eastern Screech-owls and other raptors of brodifacoum bait used for vole control in orchards. *Environmental Toxicology & Chemistry, 7*: 245–260.

Hoppe, A.H. & Krambias, A. 1984. Efficacy of three new anticoagulants against *Rattus rattus frugivorus*. *In: The organisation and practice of vertebrate pest control*, ed. by A.C.Dubock, 335–339. Haslemere, England: ICI.

Hunter, K. 1985. High-performance liquid chromatographic strategies for the determination and confirmation of anticoagulant rodenticide residues in animal tissues. *Journal of Chromatography, 321*: 255–272.

Mendenhall, V.M. & Pank, L.F. 1980. Secondary poisoning of owls by anticoagulant rodenticides. *Wildlife Society Bulletin, 8*: 311–315.

Merson, M.H., Byers, R.E. & Kaukeninen, D.E. 1984. Residues of therodenticide brodifacoum in voles and raptors after orchard treatment. *Journal of Wildlife Management, 48*: 212–216.

Newton, I., Wyllie, I. & Freestone, P. 1990. *Rodenticides in British Barn Owls. Environmental Pollution, 68*: 101–117.

Petroli, J.C. 1983. *Control of the Field Vole*. Report of the Neuchatel Canton Fishing and Hunting Service. Neuchatel.

Rammell, C.G., Hoogenboon, J.J.L., Cotter, M., Williams, J.M. & Bell, J. 1984. Brodifacoum residues in target and non-target animals following rabbit poisoning trials. *N.A. Journal of Experimental Agriculture, 12*: 107–111.

Shawyer, C.R. 1987. *The Barn Owl in the British Isles. Its past, present and future*. London, The Hawk Trust.

Thomson, W.T. 1976. *Agricultural chemicals book 3: fumigants, growth regulators, repellents and rodenticides*. California, Thomson Publications Fresno.

Townsend, M.G., Fletcher, M.R., Odam, E.M. & Stanley, P.I. 1981. An assessment of the secondary poisoning hazard of warfarin to Tawny Owls. *Journal of Wildlife Management, 45*: 242–248.

Worthing, C.R. & Walker, S.B. 1987. *The pesticide manual*. Lavenham Press, The British Crop Protection Council, Thornton Heath.

Habitat use by farmland Tawny Owls *Strix aluco*

A.R. Hardy

Hardy, A.R. 1992. Habitat use by farmland Tawny Owls *Strix aluco*. In: *The ecology and conservation of European owls*, ed. by C.A. Galbraith, I.R. Taylor and S. Percival, 55–63. Peterborough, Joint Nature Conservation Committee. (UK Nature Conservation, No. 5.)

A population of Tawny Owls was studied in farmland in Aberdeenshire where deciduous woodland forms less than 5% of the area. Owls in six isolated woods were radio-tracked to determine their nocturnal hunting ranges. Large exclusive territories were maintained up to 3.0 km in length. Utilised hunting ranges were calculated to form between 17% and 40% of the maximum home ranges. When away from their daytime roosts, owls spent most time in small core-areas of deciduous woodland, but hunted extensively over open farmland along lines of habitat.

Pellet analysis from 14 territories revealed that the principal vertebrate prey was the Field Vole *Microtus agrestis* which formed 43% by weight of the diet. Traplines in eight habitat-types indicated that the Woodmouse *Apodemus sylvaticus* is present in deciduous and mature conifer woodland and that the Field Vole occurs in immature plantation and areas of rough grassland. Movements of hunting Tawny Owls were shown to be related to the distribution of their major prey.

A.R. Hardy, Central Science Laboratory, Ministry of Agriculture, Fisheries and Food, London Road, Slough, Berkshire SL3 7HJ

Introduction

Although it is the commonest and most widely distributed owl of mainland Britain (Sharrock 1976) and is found in a wide range of habitats, the Tawny Owl *Strix aluco* has been little studied outside its preferred woodland habitat. Classic studies by Southern and his students in Wytham Wood (Southern 1970; Hirons 1985) have described the territorial behaviour, the population dynamics and the basic ecology of the bird in its optimal habitat. Numerous dietary studies based on pellet analyses reveal significant differences between owls occupying different habitats (Uttendorfer 1939; Southern 1954; Andersen 1961; Wendland 1963; Beven 1965; Glue 1971; Cheylan 1971; Smeenk 1972; Kallander 1977; Nilsson 1984; Yalden 1985; Henry & Perthuis 1986). However, few studies of feeding ecology have attempted to relate the hunting behaviour and habitat use by the owls to the distribution of their prey (Southern & Lowe 1968). Woodland Tawny Owls hunt throughout their exclusive territories, whose contiguous boundaries may be mapped from observations of calling birds (Southern & Lowe 1968). This paper describes a study of Tawny Owls in Scotland in intensively cultivated farmland surrounding small isolated woods. The extent of hunting movements of this nocturnal predator was determined by radio-tracking and hunting behaviour and diets were compared with the distribution of major prey species.

Methods

Field work was undertaken from 1972 to 1975 in 155 km² of Aberdeenshire farmland, surrounding the Ythan Valley in north-east Scotland. Pairs of Tawny Owls were initially located during autumn, when owls are most vocal, by broadcasting pre-recorded calls throughout the area to elicit responses from resident birds (Hirons 1976). Repeat censuses and the positive identification of individual owls, from unique sound spectrographs of their recorded calls, allowed adjacent pairs to be distinguished and a total count derived (Hirons 1976; Hardy 1977). This technique also permitted the remote identification of territorial adults in successive years.

Six territories were selected for intensive study. Eight Tawny Owls (six male, two female) were each fitted with a small radio transmitter attached by a nylon harness to their back, between their wings (Hardy 1977). Radio packages weighed between 4% and 7.6% of body weight and functioned for up to 120 days in the field. Resident owls, which were trapped at night in mistnets or chardonneret box traps set under roost sites, were taken to the laboratory where transmitters were fitted. After observation and testing of equipment, owls were released at their site of capture the next dusk. Owls were located with the aid of hand-held, vehicle-mounted and tower-mounted aerials. Each nocturnal radio fix involved two measurements from separate points (Heezen & Tester 1967) which were plotted onto large-scale, gridded field maps and the position of the radio-tagged owl identified at their intersection. A minimum interval of 30 minutes was chosen between subsequent fixes. Owls could be approached to within 60 m on dark nights and approximately 100 m on moonlit nights without disturbance. Occasionally hunting birds could be observed directly. Radios operated for 732 'owl-nights' during a 14 month period though the bulk of the data was collected between February and August.

Regurgitated pellets were collected at regular intervals beneath woodland roost sites in 14 territories and analysed for prey content. After fur and feathers had been removed by mechanical agitation under water (Southern 1954) bones were separated, identified using a reference collection and counted. Invertebrate exoskeletal remains were identified and sediment samples scanned for the presence of earthworm chaetae. Numbers of each vertebrate prey species consumed were converted to prey units using a conversion factor relating average individual weight to that of a 'standard 20 g mammal'; equivalent to a mouse or a vole (Southern 1954). Diet can be expressed as proportion of weight rather than numbers caught.

An extensive trapping programme was carried out during the study period to monitor small mammal populations. Eight sites, all observed to be hunted over by owls, were selected as representative of the range of habitats available within the study area. At each site, a 200 m removal trapline of ten points was operated over four consecutive nights without prebaiting. Three Longworth live traps and two breakback traps were set at each point. Seven trapping sessions were conducted at two-monthly intervals followed by a final session after an interval of six months.

Results

The distribution of Tawny Owls in the 155 km^2 study area depended heavily on the patchiness of woodland which formed less than 5% of the total area. 26 separate pairs were located by census, of which 18 were associated with isolated woods while the remaining eight were located on two larger estates. All suitable farms, copses, plantations and more extensive woodland areas were checked repeatedly during the study period and it is unlikely that any pair was missed. Resident Tawny Owls in autumn and winter were only heard to hoot territorially from perches in trees or on buildings. Although birds could be attracted to fly low over a pre-recorded call played in open farmland between woods, the owls invariably flew to a tree perch before responding

vocally. Consequently the mapping technique for owls in woodland (Southern & Lowe 1968) was found to be inadequate outside two larger wooded estates. Radio-tracking was therefore used to elucidate hunting movements outside woodland. Six territories were selected for intensive study, based on five isolated woods varying in size from 0.9 to 7.8 ha and the remaining one on the edge of a more extensive wooded estate at Auchmacoy.

Hunting ranges

Radio-tracking data may be presented and analysed in many different ways. Several calculations of hunting ranges were made from the data on locations of active owls not at roost. For example, a typical map of radio-fixes from the tagged pair of Tawny Owls at Macharmuir is presented as the total number of fixes per hectare square (Figure 1a). The most striking features of this distribution, as shown by all the radio-marked owls, are the discontinuity, the importance of relatively few grid squares and the long range length. Examination of Figure 2 shows more clearly for the same pair the importance of a few grid squares within the observed range of the owls. All marked birds showed the same spatial pattern of movements. If individual point radio-locations are plotted, the maximum range length (Stickel 1954) and maximum range area (Odum & Kuenzler 1955) may be determined (Figure 1b). The farmland ranges of Tawny Owls in this study were larger than published figures and were found to extend up to 3 km in length (Table 1).

The spatial discontinuity shown in Figure 1 is a function of both the measurement technique and the owls' behaviour. The Tawny Owl hunts from a perch and all fixes were measured for perched birds. Although the maximum home range is relevant to the spacing of adjacent pairs of owls through the study area, this polygon includes large areas unused by the owls when hunting (Forbes & Warner 1974). The owls appeared to use corridors when moving around their ranges as they hunted along lines of structural perches. Therefore a more objective measure of hunting range was derived in this study as being of greater biological relevance to the interpretation of hunting behaviour.

Utilised hunting ranges were therefore calculated by measuring the total areas of habitats suitable for small mammals (their major prey) out to the external points of all locations along the corridor used by the owls. This was based on three assumptions: (1) in open farmland small mammals are largely restricted to within 20 m of field edges, (2) a perched Tawny Owl is assumed to be hunting throughout the period

Table 1.

Site	Sex	Fixes	Maximum Range (ha)		Hunting Range Size (ha)	
			Length	Area	Core	MUR
Macharmuir	m,f	130	1.6	105.8	1.2	43.3
Tarty	m,f	170	2.7	72.5	7.6	29.8
Waterside	m	110	3.0	185.9	5.6	31.7
Auchmacoy	m	53	1.3	63.1	2.5	21.0
Cross Stone	m	44	0.8	19.8	6.7	17.7
Knockhall	m	41	2.2	43.3	0.9	15.6
Totals	6m,2f	548				

Maximum range length (Stickel 1954) and maximum range areas (Odum & Keunzler 1955) for eight radio-tracked Tawny Owls in six farmland territories between July 1974 and August 1975. Estimates of core-areas and minimum utilised ranges (MUR) are included as defined in the text.

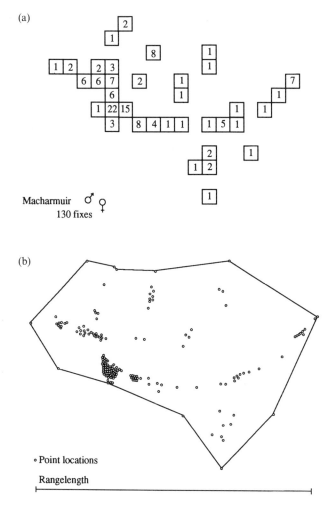

(a)

Macharmuir ♂ ♀
130 fixes

(b)

○ Point locations

Rangelength

Figure 1.   Distribution of radio-fixes for radio-tracked male and female Tawny Owls from February to August 1975 at Marcharmuir. Fixes were taken at a minimum interval of 30 minutes on active owls not at roost. Data are expressed as fixes per ha grid square (Figure 1a) and as point locations to show maximum home range and rangelength (Figure 1b).

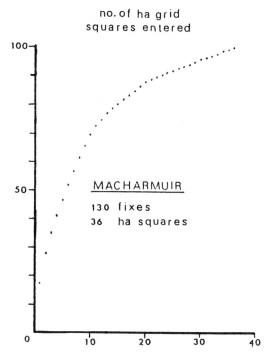

Figure 2.   Cumulative percentage of radio-fixes in each grid square entered by radio-marked male and female Tawny Owls at Marcharmuir. Squares are ranked from left to right in decreasing order of occupation.

that it is away from the roost, and (3) an owl is unlikely to detect small vertebrate prey at distances greater than 20 m when perched on fence posts or walls at a height of approximately 1 m above the ground. From the distribution of the radio-fixes in each monitored territory, areas hunted over by the owls were calculated as corridors 20 m either side of several distinct habitat types. For example, a 20 m strip was added to the circumference of all plantations, woods and farm steadings used, while a 40 m corridor was applied to all fencelines, dykes and hedgerows (Figure 3). Calculated corridor areas were then added to those of woodland, plantation and farm steading to produce a minimum estimate of the utilised hunting range (MUR).

The concept of core-area (Kaufman 1962) is applied to that area of woodland, wooded garden or farm-steading which forms the focus of the owls' activity

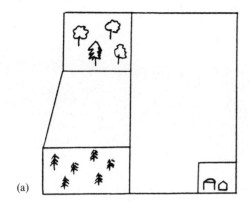

(a)

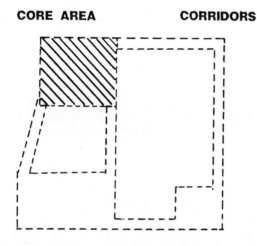

CORE AREA   CORRIDORS

(b)   MINIMUM UTILISED RANGE   (M.U.R.)

Figure 3.   Diagrammatic representation of hypothetical home range including woodland, plantation and farm buildings (Figure 3a). The calculated minimum utilised range (MUR) is illustrated (Figure 3b).

as limited by the structural boundary of the particular habitat. Here they roost, nest, defend the area vocally against intruders and may spend a large proportion of the total hunting time.

At the three sites with the most data, estimates of minimum utilised ranges (MUR, Table 1) vary from 41% to only 17% of the maximum home range. Despite hunting widely over open habitats of rough grass bordering burns, fences and field edges which accounted for up to 70% of the MUR, the owls spent a disproportionate time in their core-areas (Figure 2). Whereas the combined core areas of all 6 sites formed only 15.4% of the total MURs, the radio-marked birds were located within their boundaries in 54% of all fixes (Figure 4). Data for Tarty, Waterside and Macharmuir suggest a trend where the larger the core-area the smaller the MUR used by the owls. In the Cross Stone territory, with the largest single core-area, the male was recorded outside mature conifer woodland in only 23% of

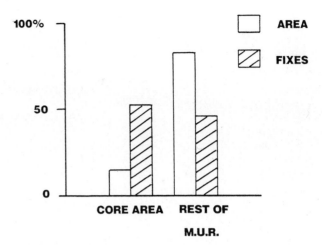

Figure 4.   Distribution of radio-fixes between core-areas and remainder of minimum utilised range (MUR). The data from all six territories with radio-marked Tawny Owls are combined.

fixes and exhibited the smallest range length of any radio-marked owl (Table 1). At each fix, the habitat was recorded in which the owl was located and the combined data for all sites was used to determine the habitat preference. It is clear that Tawny Owls strongly selected deciduous woodland (Figure 5) but were also found on, or in, farm buildings, wooded gardens, immature plantations, including strip plantations and a variety of open perches around field edges.

Typical movements around the hunting range are illustrated by a male Tawny Owl tracked from February to August at Waterside (territory 1 in Figure 6). During this period he hunted to supply food for the female and subsequently a brood of three chicks. A mixed wood overlooking the estuary formed the core-area in which the pair roosted and nested and which, together with adjacent open scrub, provided 51% of all fixes. The recorded hunting range extended from a hedgerow west of the river, north-east to some farm buildings, a linear distance of

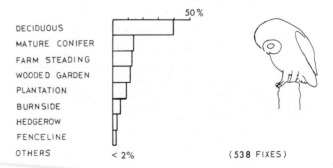

Figure 5.   Distribution of active radio-marked Tawny Owls according to the habitat in which they were located. This is based on 538 radio-fixes combined for all six territories.

58

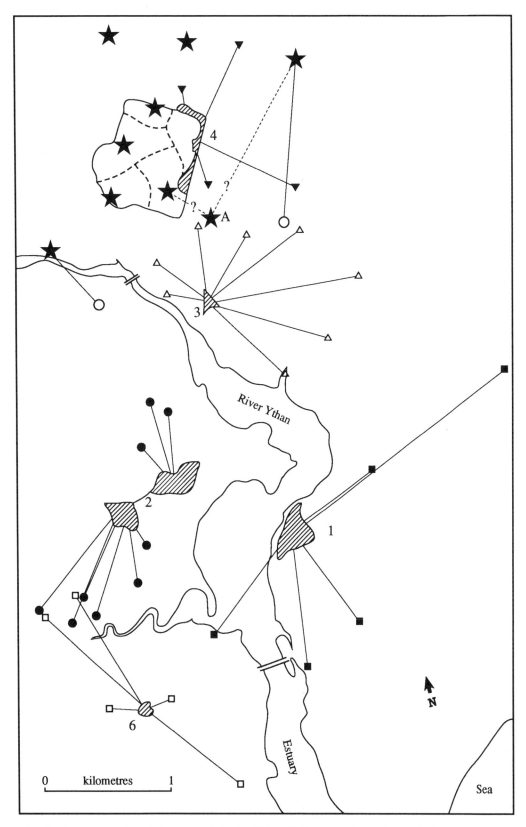

Figure 6. Distribution of 13 pairs of Tawny Owls in the Ythan Valley. For five radio-tracked territories, each symbol represents the furthest location in any direction connected to the core-area of the owl(s) involved. Closed stars indicate an occupied territory where resident owls responded to tape recorded calls. Open circles denote additional woods from which owls of a given territory responded to recorded calls. Approximate territory boundaries are shown within a continuously wooded estate. An adult of unknown origin was trapped at point A.

some 3.0 km. The male regularly hunted at Waterside Farm mostly within a large storage barn containing potatoes, adjacent to the core-area of woodland. Hunting along rough heath bordering a burn was recorded on several occasions in both the southern and northern parts of the range, extending onto the Sands of Forvie National Nature Reserve.

Both male and female Tawny Owls were radio-marked at two sites but no evidence was found of differential use of the territory by the two sexes. The distribution of 13 pairs of Tawny Owl is shown in Figure 6 to include five of the radio-tracked areas. As in continous woodland (Southern & Lowe 1968; Southern 1970) each territory was found to be hunted exclusively by a particular pair of owls. Only one case of overlap was recorded, where neighbouring owls hunted around a set of farm buildings, but six months apart (territories 2 and 6).

Hunting behaviour

Primarily nocturnal, the Tawny Owl hunts from a perch dropping onto vertebrate prey below (Southern & Lowe 1968). Comparison with measurements of other British owl species reveals that the Tawny is well adapted to woodland. The short broad wings, a high wingloading and low aspect ratio (wingspan divided by wing length) which increases drag in flight allow great manoeuvrability among trees.

At dusk owls leave their diurnal roost, calling from the site before moving from perch to perch, actively hunting. If unsuccessful from one perch, the owl will fly to another after a few minutes. Radio tracking and observation confirmed that this first foray of the night extends for about one hour, during which owls proved unresponsive to tape-recorded calls used in census and very few unprovoked calls were heard (Hansen 1952). The same hunting technique was used for vertebrate prey in open habitats where the owl perches on available fenceposts and walls. No evidence was found in this study of Tawny Owls hunting on the wing by repeated quartering of open ground (Nilsson 1978). When feeding on earthworms and other invertebrates, the Tawny Owls perched on the ground to pick them off the surface (personal observation; MacDonald 1976). This behaviour was observed frequently in wet weather when radio-marked owls ran about on the ground in grass fields within a few metres of the margins. They were rarely located hunting on open ground further than about 20m from the field boundaries.

Changes in daylength affect the length of activity away from roost. As nights decreased in length during the summer, Tawny Owls left roost progressively earlier in relation to sunset (Figure 7). A more varied return to roost around dawn was probably affected by weather variables which in turn affect hunting success. This variability was seen at its extreme on midsummer's night (5.5 hours of twilight) when the male at Auchmacoy was active, away from roost, for more than 10 hours. During this study, daytime hunting was regularly recorded for several Tawny Owls during the breeding season when food demands were at their peak in midsummer.

Diet

Pellet samples were obtained from 14 territories that varied in habitat from continuous woodland to isolated woods and immature plantation. This sample included the six territories occupied by radio-marked owls. A total of 580 pellets contained 922 vertebrate prey equivalent to 1,053 prey units. Small mammals formed 92% of the number found, but only 70% of the diet after correction for weight (Figure 8). Field Voles *Microtus agrestis* were the principal prey (43%) while the two major woodland species, Woodmouse *Apodemus sylvaticus* and Bank Vole *Clethrionomys glareolus*, together formed only 17% by weight. Larger mammals, Rats *Rattus norvegicus* and Rabbit *Oryctolagus cuniculus* accounted for 19.5%. Starlings *Sturnus vulgaris* were the most important single constituent (55%) of the bird component that formed the remaining 10% of the vertebrate diet.

Seasonal variation in the diet was seen in the pellet contents during the period January to May. While the total proportion by weight of small mammals fell from 78.5% to 56.7%, large mammals rose from 10.7% to 33.1%. Within this overall pattern, Field Voles decreased from 45.6% to 33%, Woodmice increased from 7.6% to 13.3% and Rabbits increased in importance in the diet from 2.4% to 23.3%. Though the overall proportion of birds in the diet remained constant (10.8% to 9.8%) Starlings increased in importance. Direct observation of adult owls bringing prey back to active nests, supplemented with data from surplus prey remains in the nests, reflected this change in diet. Out of 693 prey units recorded at the nests, Rabbits formed 28.9%. In this sample, birds formed overall 44.2% of prey units with Starlings as the most important single species (23.7%). The spectrum of bird species identified at the nests was broader than in the pellets ranging from Mallard duckling *Anas platyrhyncos* and juvenile Redshank *Tringa totanus* to two adult Woodpigeons *Columba palumbus*.

A fifth of all pellets consisted of characteristic fibrous pellets (Southern 1954) which on microscopic examination revealed large numbers of earthworm

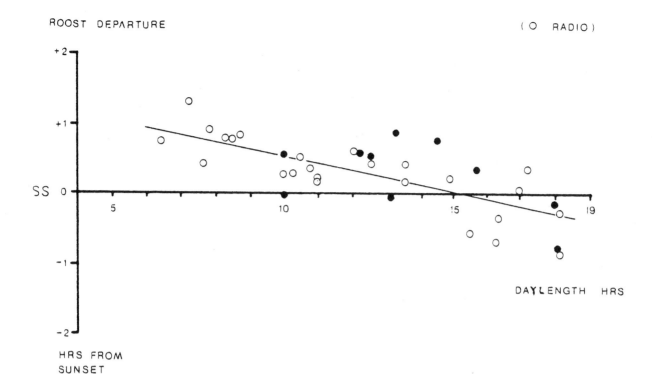

ROOST DEPARTURE   (O RADIO)

SS

DAYLENGTH HRS

HRS FROM
SUNSET

Figure 7.   Onset of evening activity of Tawny Owls, or time of departure from roost as measured from sunset (SS), in relation to changes in day length. The symbols denote radio-determinations (○) and observations (●); $y = 1.57 - 0.1x$, $r = 0.728$, $p < 0.001$.

chaetae. Significantly more earthworm pellets were found during the wetter winter months, December to March, than during the breeding season, April to June. Remains of other invertebrates found less frequently included ground beetles and the head capsules of Lepidoptera larvae. These were found principally between March and May, concurrent with the decrease in small mammal content of the diet in the early breeding season.

Trapping results

The species composition of the resident small mammal population at each site may be compared by the frequency of capture (Linn 1954) expressed as a percentage of the total catch summed over the 8 trapping sessions (Figure 9). In mixed woodland, (mature conifer stands and isolated woods), woodmice predominated (A, B, E and C). In contrast the Field Vole together with the Common Shrew *Sorex araneus* were the major species found in immature plantation and rough heathland where the field layer of vegetation was very dense (F and G). A hedgerow and ditch supported almost equal numbers of the four main species trapped (D). The lowest density of small mammals was found on heather moorland on the Sands of Forvie National Nature Reserve where shrews (*S. araneus* and *S. minutus*) were the commonest species (H).

% PREY UNITS

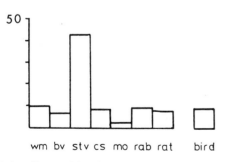

Figure 8.   Composition by weight of vertebrate content of Tawny Owl diet determined by analysis of 574 pellets from 14 territories, including the six containing radio-marked owls. Individual prey species are: wm, *Apodemus sylvaticus*; bv, *Clethrionomys glareolus*; stv, *Microtus agrestis*; cs, *Sorex araneus*; mo, *Talpa europaea*; rab, *Oryctolagus cuniculus*; rat, *Rattus norvegicus*.

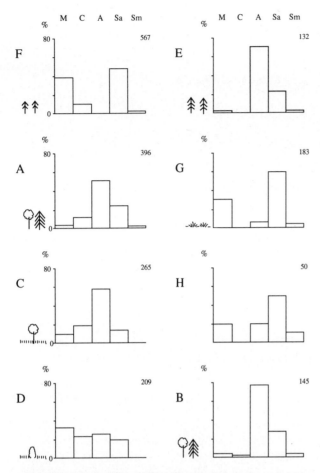

Figure 9. Composition of small mammal catches over 21 months of eight removal traplines (sample sizes shown). These were run simultaneously in different habitats: A and B, mixed woodland; C, a small isolated wood; D, a hedgerow; E, mature conifers; F, immature plantation; G, rough heathland; H, heather moorland. The species trapped were: M, *Microtus agrestis*; C, *Clethrionomys glareolus*; A, *Apodemus sylvaticus*; Sa, *Sorex araneus*; Sm, *Sorex minutus*. Total number of small mammals caught during all trapping sessions are shown.

Discussion

Farmland Tawny Owls in this study occupied large, exclusive territories with long range lengths. In contrast other studies have found much smaller territories, 12 to 24 ha (range lengths 300 m to 950 m), in continuous woodland in Wytham Wood (Southern 1970; Hirons 1985), while owls in more open parkland occupied territories ranging from 20 to 32 ha in total area (range lengths 600 m to 1200 m, Southern 1970). The maximum recorded range size in the present study was 15 times greater than the smallest viable territory in closed woodland at Wytham (Southern 1970). In terms of areas utilised by the hunting owls, the Aberdeenshire figures are slightly larger than Southern's parkland figures and of the same order as in open parkland in Denmark (Andersen 1961: 30 to 50 ha in area, range

lengths 1.5 to 2.2 km). Hirons (1985) working at the same time in Warwickshire farmland found that defended areas (equivalent to the maximum home ranges of this study) showed large variation with a mean size of 37 ha.

The overall territory size depends on the patchiness and distribution of different habitats within it. In this study, owls with small core areas appeared to compensate by hunting over larger areas outside woodland (Table 1). Hirons (1985) found an inverse correlation between defended area and the area of closed woodland. However, there was no relationship between defended area and an estimate of the available foraging area which appeared relatively constant over a threefold variation in total territory size. It should be noted that although the minimum utilised range (MUR) probably underestimates total area usage, since it is related to vertebrate prey distribution, no instances were found in this study of radio-marked owls feeding on earthworms away from field edges. The assumption that small mammals are largely limited to within 20 m of field edges was based on trapping studies in farmland which clearly indicate that voles are normally restricted to hedgerows and the edges of fields and woodland (Kikkawa 1964; Pollard & Relton 1970; Eldridge 1971; Jefferies *et al.* 1973).

Although the Woodmouse may range widely from the field edges according to crop types, time of year and the proximity of woodland (Kikkawa 1964; Pollard & Relton 1970; Jefferies *et al.* 1973) or may live within certain field crops all year round (Green 1979), this species formed only 10% of the diet of the Tawny Owl in this study while voles formed 50%.

The demonstration of exclusive territories in open farmland raises the behavioural question of how a pair of Tawny Owls may defend a territory up to 3 km in length. Occupation of isolated woodlots is regularly advertised over long distances using territorial calls in the autumn and early winter. Lack of overlap between neighbouring hunting birds observed during the study period may reflect the stability of territory boundaries (Southern 1970) and the low annual replacement rate of resident owls (20%-Southern 1970; 11%-Hardy 1977). The only recorded boundary dispute outside closed woodland involved one pair and a neighbouring male in a shelter belt between two core-areas and resulted in an aerial chase after 30 minutes of vocal conflict.

Deciduous woodland was strongly selected by owls in this study and radio-marked birds spent most time there. However, pellet analysis revealed that woodland species of small mammals formed only

17% by weight of the diet. This contrasts with owls of more closed woodland at Wytham Wood where Woodmice and Bank Voles formed 46% of the diet (Southern 1954) and accordingly Field Voles formed only 10%. In another farmland study, birds were found to be the major prey category (34%) of Dutch Tawny Owls (Smeenk 1972); Field Voles forming only 11% of their diet.

Small mammal trapping in the current study confirmed that mice are largely confined to woodland and that the Field Vole is the major species in immature conifer plantation and rough grassland around burns, hedges and field edges. Radio-tracking revealed extensive hunting movements out of woodland and into adjacent farmland where Tawny Owls preyed largely on the Field Vole, the predominant rodent in open habitats and the most widely distributed small mammal in the study area.

The conclusion to be drawn from the many dietary studies of the Tawny Owl which have been carried out from woodland to urban habitats and the few studies, including this one, which have related prey distribution to the owls' foraging behaviour is that this species is primarily an opportunist predator. Using a sit and wait technique, the Tawny Owl preys on those species most available in the habitats within its hunting range. Territory size is considered to be inversely related to prey abundance. There remains considerable scope for further comparative work on the feeding ecology of the Tawny Owl outside woodland, not only in open habitats with poor tree cover but particularly in urban areas where even less is known about territorial distribution and habitat use.

Acknowledgements

The Natural Environment Research Council supported this work and I thank George Dunnet for his encouragement and advice throughout. I am particularly grateful to the late Mick Southern for his inspiration, advice and critical discussion of work with Tawny Owls. Thanks are also due to the many farmers and landowners without whose co-operation this study would not have been possible.

References

Andersen, J. 1961. En nordsjaellandsk Natugle-bestand *Strix aluco* yngletiden. *Dansk Ornitologisk Forenings Tidsskrift, 55*: 1–55.

Beven, G. 1965. The food of Tawny Owls in London. *London Bird Reports, 29*: 56–72.

Cheylan, G. 1971. La regime de la chouette-hulotte (*Strix aluco*) a salerres (Vor). *Alauda, 39*: 150–155.

Eldridge, J. 1971. Some observations on the dispersion of small mammals in hedgerows. *Journal of Zoology London, 165*: 530–534.

Forbes, J.E. & Warner, D.W. 1974. Behaviour of a radio-tagged Saw-whet Owl. *Auk, 91*: 783–795.

Glue, D.E. 1971. Avian predator pellet analysis and the mammalogist. *Mammal Review, 1*: 53–62.

Green, R. 1979. The ecology of Wood Mice (*Apodemus sylvaticus*) on arable farmland. *Journal of Zoology London, 188*: 357–377.

Hansen, L. 1952. (The diurnal and annual rhythm of the Tawny Owl.) *Dansk Ornitologisk Forenings Tidsskrift, 46*: 158–172.

Hardy, A.R. 1977. *Hunting ranges and feeding ecology of owls in farmland.* Unpublished. PhD. thesis, Aberdeen University.

Heezen, K.L. & Tester, J.R. 1967. Evaluation of radio-tracking by triangulation with special reference to deer movements. *Journal of Wildlife Management, 31*: 124–141.

Henry, C. & Perthuis, A. 1986. The feeding regime of the tawny owl *Strix aluco* L. in the forested areas of central France. *Alauda, 54* (1): 49–65.

Hirons, G.J.M. 1976. *A population study of the Tawny Owl (Strix aluco) and its main prey species in woodland.* Unpublished. D. Phil Thesis, University of Oxford.

Hirons, G.J.M. 1985. The effects of territorial behaviour on the stability and dispersion of Tawny Owl (*Strix aluco*) populations. *Journal of Zoology, London, (B) 1*: 21–48.

Jefferies, D.J., Stainsby, B. & French, M.C. 1973. The ecology of small mammals in arable fields drilled with winter wheat and the increase in dieldrin and mercury residues. *Journal of Zoology, London, 171*: 513–539.

Kallander, H. 1977. Kattuglans *Strix aluco* och hornugglans *Asio otus* bytesval vid Kvismaren-en jamforelse. *Vår Fågelvärld, 36*: 134–142.

Kaufman, J.F. 1962. Ecology and social behaviour of the coati, *Nasua narica*, on Barro Colorado Island, Panama. *University of California Publications in Zoology, 60*: 95–222.

Kikkawa, J. 1964. Movement, activity and distribution of small rodents *Clethrionomys glareolus* and *Apodemus sylvaticus* in woodland. *Journal of Avian Ecology, 33*: 259–299.

Linn, I. 1954. Some Norwegian small mammal faunas: a study based on trapping in West and North Norway. *Oikos, 5*: 1–24.

MacDonald, D.W. 1976. Nocturnal observations of Tawny Owls *Strix aluco* preying on earthworms. *Ibis, 118*(4): 579–580.

Nilsson, I.N. 1978. Hunting in flight by Tawny Owls *Strix aluco*. *Ibis, 120* (4): 528–531.

Nilsson, I.N. 1984. Prey weight, food overlap, and reproductive output of potentially competing long-eared and tawny owls. *Ornis Scandinavica, 15*(3): 176–182.

Odum, E.P. & Kuenzler, E.J. 1955. Measurement of territory size and home range size in birds. *Auk, 72*: 128–137.

Pollard, E. & Relton, J. 1970. A study of small mammals in hedges and cultivated fields. *Journal of Animal Ecology, 7*: 549–557.

Sharrock, J.T.R. 1976. *The atlas of breeding birds in Britain and Ireland.* Tring, British Trust for Ornithology.

Smeenk, C. 1972. Okologische vergleiche zwischen Waldkauz und Waldohreule. *Ardea, 60* (1–2): 1–71.

Southern, H.N. 1954. Tawny Owls and their prey. *Ibis, 96*: 384–410.

Southern, H.N. 1970. Natural control of a population of Tawny Owls. *Journal of Zoology, London, 162*: 197–285.

Southern, H.N. & Lowe, V.P. 1968. The pattern of distribution of prey and predation in Tawny Owl territories. *Journal of Animal Ecology, 37*: 75–97.

Stickel, L.F. 1954. Home ranges in small mammals. *Journal of Mammalogy, 35*:1–15.

Uttendorfer, O. 1939. *Die Ernahrung der deutschen Raubvogel und Eulen und ihr Bedeutung in der heimischen Natur.* Berlin, J. Neumann-Neudamm.

Wendland, V. 1963. Funfjahrige Beobachtungen an einer Population des Waldkauzes *Strix aluco* in Berliner Grunewald. *J. Orn. Lpz, 104*: 23–47.

Yalden, D.W. 1985. Dietary separation of owls in the Peak District. *Bird Study, 32*: 122–131.

Population ecology of Little Owls *Athene noctua* in Central Europe: a review

K.M. Exo

Exo, K.M. 1992. Population ecology of Little Owls Athene noctua in Central Europe: a review. *In: The ecology and conservation of European owls*, ed. by C.A. Galbraith, I.R. Taylor and S Percival, 64-75. Peterborough, Joint Nature Conservation Committee. (UK Nature Conservation, No.5.)

During the last twenty years many studies on the ecology of Little Owls *Athene noctua* have been carried out in Central Europe, especially in Germany and Switzerland. This paper reviews the results of several published and unpublished long-term studies and examines the various factors that limit and regulate European Little Owl populations. Short summaries are given for the following topics: (a) the present situation and population density in some European countries; (b) habitat structure and nest-site ecology; (c) home range and territory size; (d) the annual timing of breeding; (e) clutch size; (f) breeding success; and (g) the mortality rates of adult and juvenile birds. On the basis of these data the following are discussed: (a) if the reproduction rate is able to compensate for mortality, the population decrease over wide areas of Central Europe is the result of low reproduction rates; (b) the role of habitat structure, especially of nest-site availability in limiting breeding density; and (c) density dependent regulating processes.

Differences in population density within and between different areas in Central Europe are largely related to nest-site availability. The occurrence of old pollarded trees and orchards, and the availability of suitable hunting grounds, in particular grassland habitats with short vegetation throughout the year, are especially important. Nest-site availability appears to be the 'ultimate' limiting factor over much of Europe today. For some regions we have to assume that mortality, especially winter mortality, limits population density. In more densely populated areas, territorial behaviour acts as a 'proximate' regulating factor.

Klaus-Michael Exo, Institute fur Vogelforshung, Vogelwarte Helgoland, An der Vogelwarte 21, 2940 Wilhelmshaven-Rustersiel, Germany

Introduction

The Little Owl *Athene noctua* belongs to the Turkestanian-Mediterranean fauna and has a trans-Palaearctic distribution (Voous 1960). Primary habitats are savannas, maquis, dry hilly steppes, semi-deserts and scree regions of dry temperate zones. The breeding range of the Little Owl spread to central Europe up to the July isotherm of 17°C. The northern as well as the altitudinal range boundary may be caused by climate. In Europe, Little Owls have colonized a wide range of habitats with the exception of uplands (> 300 m) and woodlands. Permanent grasslands which have short vegetation (< 15 cm) throughout the year, in particular pastures flanked by lines of pollarded trees (mostly *Salix* spp.) and parkland well endowed with hedgerow trees and old orchards, affording ample nest-holes, provide optimal habitats.

The population of Little Owls have declined over much of Europe since c. 1950, with the possible exception of the Mediterranean region, (Table 1; e.g. Kesteloot 1977; Juillard 1989). According to Glutz & Bauer (1980) the widespread decline is more pronounced than for any other European owl species. Recent declines are primarily due to habitat changes associated with the intensification and mechanisation of agriculture. These changes in agricultural practices have led to a decline in the total area of grassland habitats and the clearing of old orchards and pollarded trees. This results in a dramatic reduction in the avaiability of hunting grounds and the number of suitable nest-sites.

Knowledge of spatial and temporal variations in breeding densities of owls is poor. Owl populations, like other bird species, may be limited by scarcity of resources, such as nest-sites or food. On the other hand, density may be suppressed below the resource level by excessive mortality or unsuccessful reproduction (Newton 1979). For Little Owl populations in Central Europe Juillard (1989) assumed that the density was supressed below the resource level. This is based on the assumption that Little Owls do not produce enough young to compensate mortality (but see Exo & Hennes 1980; Exo 1983). In the absence of human influences most owl or raptor populations are limited by food supply in winter or, in summer, by food supply or nest-sites, which ever is scarcest (Newton 1979). If the limiting factor is removed, population density increases until the population reaches the new limit set by the same or another factor. Some factors act as 'ultimate' limiting factors whilst others are 'proximate' factors. Ultimate factors are those which directly affect the population, such as food and nest-sites, and proximate factors, mediate the effects of ultimate limiting factors, for example territorial behaviour.

During the last two decades many studies on the ecology of Little Owls have been published, especially in Germany and Switzerland. The aim of this paper is to summarize recently published and unpublished data on factors limiting and regulating populations. Population density, territory size, breeding biology and mortality are focused on and discussed in terms of their relevance to obtaining an understanding of factors which affect population regulation. Although natal dispersal and adult nest-site fidelity may also be important regulating factors no allowance is made for these since they are topics dealt with in another paper (Exo in prep.). When dealing with Little Owl population ecology one should always keep in mind that all habitats colonised in central Europe are so-called secondary habitats; primary habitats are steppes and semi-deserts. Many of my conclusions are tentative. For methodological details readers should refer to the cited literature.

Results and discussion

Population density

Table 1 gives an overview of the number of Little Owls in some European countries. Habitat destruction has resulted in the distribution becoming more disjunct. In suitable habitats or in areas with artificial nest-sites Little Owls breed in 'clumps', the scale of clumping varying from area to area. Study areas were seldom chosen at random, most observers favouring studies in densely populated areas. It is therefore meaningless to calculate the overall mean density for large geographical regions unless the proportion of habitat types examined or the density of owls in each can be given.

The mean density of Little Owls is seldom above 0.3–0.5 pairs/km² (Table 2) over areas of at least 100 km² (Village 1984; Kostrzewa 1988). This applies also to areas of 'optimal' habitats in parts of North Rhine-Westfalia and the Netherlands (Petzold & Raus 1973; Hegger 1977; AG Zum Schutz Bedrohter Eulen NRW 1978; Glutz & Bauer 1980; Mildenberger 1984; Loske 1986; Illner 1988, Kampfer & Lederer 1988). With the exception of the Maas-Waal basin near Nijmegen (Table 2 and Visser 1977) higher densities only occur on a local scale, especially with extensive orchards or high numbers of pollarded trees. For example, in grassland areas with pollarded trees such as the Lower Rhine, Middle Westfalia and Julicher Borde [edges of villages] (Glutz & Bauer 1980) and or where extensive orchards exist in areas such as Sundgau/Alsace (Kempf 1973). Between 1–2 pairs/km² were found in Sundgau/Alsace, with clumps up to 4–6 pairs/km².

Glutz (1962) and Visser (1977) reported extremely high concentrations: 15–20 breeding pairs/km². Such breeding densities can only be proved by nest-searching and ringing. Counts based on playback experiments (Exo & Hennes 1978) or records from sightings only are not sufficient to verify such values. In general nearest distances between neighbouring breeding holes were about 100–150 metres (Ullrich 1973; Knotzsch 1978; Exo 1983; Epple & Holzinger 1987), although in exceptional cases distances between neighbours were as low as 50 metres (Glutz 1962).

Home range and territory size

Information on the size and utilisation of the home range was obtained from two radio telemetry studies on the Lower Rhine, one of the most densely populated areas of Little Owls in central Europe (Exo 1983). The home range sizes of 19 radio tracked males varied from 2 to 107 ha (annual mean = 14.6ha, Finck 1988) and of six neighbouring pairs between 1 and 50 ha (annual mean = 14.5 ha, Exo

Table 1.   Estimated number of Little Owl pairs in some European countries

Country	Year	No. of pairs	Reference
Spain	?	> 50,000	Cramp & Simmons 1985
France	1970–75	30,000–80,000	Yeatmann 1976
Belgium	1950	± 12,000	Kesteloot 1977
	1972	± 4,000	
Netherlands	1979–85	8,000–12,000	SOVON 1987
West Germany	1982–84	5,000–6,000	According to Witt 1986)
East Germany	1985–86	250–300	Schonn 1986
	1987–88	110–180	Schonn et al. 1991
Switzerland	1950	± 800	Juillard 1989
	1980	185	
	1986	125	
Britain	1968–72	7,000–14,000	Sharrock 1976

Table 2. Little Owl breeding density in different regions. Only investigations based on area sizes of > 100 km² are considered.
A – arable land, F – mixed farmland, G – grassland, mA/mG – mainly arable land/grassland

Country & Locality	Habitat	Density (pairs/km²)	Range	Area Size (km²)	Study period	References
West Germany						
Schleswig-Holstein						
Westensee, Preetz	10–20%	0		250	1974/75/77/78	Ziesemer 1981
Norderstedt	F	0.2		108	1978	Ziesemer 1981
Bergenhusen	77%G	0.1		100	1975–78	Ziesemer 1981
Nordrhein-Westfalen						
Soest	mG	0.7	0.5–1.0	200	1971/72	Petzold & Raus 1973
Soest	mG	0.1	<0.1	300	1971/72	Petzold & Raus 1973
Werl	60% A	0.3	0.2–0.5	125	1974–86	
Lippstadt	F	0.3–0.4		240	1976–87	Kampfer & Lederer 1988
Viersen, Kempen	F	0.6		105	1976	Hegger 1977
Netherlands	F	1.7	0.7–17	473	1974–76	Visser 1977
Nijmegen						
Switzerland,	F	1	<1.8	275	1973–80	Juillard 1984
Ajoie						
N & E France	F	0.1		c.1100	1986/87	Genot 1988

1987). Territory sizes of the 19 tracked males ranged from 1 to 68 ha (mean = 12.3 ha, Finck 1988). The average size of the territories differed between seasons. Territory size was greatest during the main period of courtship, in March (mean = 28.1 ha, n = 7) and smallest during the period in which young were reared and at the beginning of the moult period, in July/August (1–4 ha, mean = 1.6 ha, n = 6). Territory size was also influenced by the duration of settling in a particular territory. Birds which were not established (new settlers without breeding experience in a particular territory) occupied on average bigger territories than established birds (birds with breeding experience in the concerning territory); (mean = 19.4 ha, n = 17 and mean = 6.2 ha, n = 19 respectively; Finck 1988). The territories of neighbouring pairs did not overlap throughout the year. The overlap shown on the map (Figure 1) was mainly due to plotting the maximum polygon area.

The recorded territory sizes corresponded well with population density in optimal habitats of central Europe, where clumps of between 4 and 6 pairs/km² were found. During the main period of courtship Little Owls occupied territories of about 28 ha with a mean annual territory size of approximately 15 ha. Based on the assumption of optimal division of available space this results in an abundance of 3.6 and 6.7 pairs/km² respectively. Thus, the distribution of Little Owls depends on the distribution pattern of the available nest-sites. If settlement of vacant territories happens asynchronously then the theoretical calculated values are only reached in exceptional circumstances (Maynard Smith 1974; Davies 1978). Future studies must show to what extent territory size varies with food abundance and

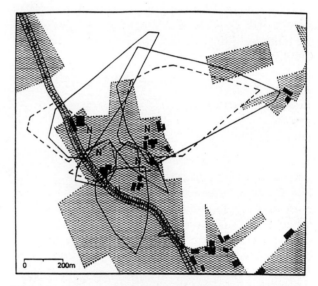

Figure 1. Home ranges (maximum polygon) of five pairs of Little Owls tracked from February to April 1984 on the Lower Rhine. Continous lines – males, broken lines = females. N = nest sites. Stippled areas show permanent grassland, black indicates farmhouses.

if territory size is influenced by population density (Huxley 1934; Tompa 1962; Southern 1970).

Timing of breeding

In Central Europe most Little Owls began laying between early April and mid-May. The majority of Little Owls laid during the last two weeks of April (Labitte 1951; Ullrich 1973, 1980; Knotzsch 1978; Glue & Scott 1980, Schonn 1986). In years with unfavourable weather conditions, during late winter or early spring, egg laying was delayed by a period of between 1 and 2 weeks. This was especially true for years with unusually long and snowy winter periods (Exo 1983, 1987; Finck 1988). Figure 2 presents typical examples of the annual and year-to-year variations. In the Lower Rhine study area there was a positive correlation between the number of days with a snow cover of more than 1 cm in January or February and the hatching dates (Exo in prep.). Illner (1979) and Ullrich (1980) observed that egg laying started earlier in peak vole years than in low vole years (Figure 2). No relationship between laying date and the age of the female has yet been shown (Ullrich 1980).

Estimates of the length of the incubation period varied between 22 and 28 days which means that most young hatched during the last week of May. The nestling period lasted between 30 and 35 days (Haverschmidt 1946; Ullrich 1973; Juillard 1979). The

young left the nest-holes between mid-June and the end of July, dispersing when they were between 2 and 3 months old, soon after they become independent. Independence is usually reached between the end of July and mid-September (Exo & Hennes 1980; Knotzsch 1978; Ullrich 1980).

Clutch size

Clutches of 1–10 eggs have been recorded but most clutches contain 3–5 eggs (Table 3, Figure3 and 4). As for many other European birds the clutch size becomes larger from a west to east direction (Figure 3), (Lack 1947; Klomp 1970; Murray 1976). The geographic trend in the west-east direction is assumed to be related to climate and the resulting effect that this has on food abundance. According to Klomp (1970) prey density, especially insect density, is higher in the warmer and drier areas of central Europe than in the oceanic influenced western parts. Furthermore, although in many species the average clutch size increases with latitude (Lack 1947/48; Klomp 1970; Murray 1976), in European Little Owls a reversed trend appeared, clutch size decreased from south to north (Figure 3). This trend can also be related to climatic differences. Little Owls favour warm and semi-arid conditions.

Several long-term studies give evidence that clutch size, as in most owl and raptor species, varies in accordance with food abundance, especially with vole density, in the pre-laying period (Knotzsch 1978; Illner 1979; Ullrich 1980; Exo 1983). For example on the northern foot-hills of the Swabian Alb in 1975, a year when Common Voles *Microtus arvalis* were abundant, the mean number of eggs per clutch was 4.6 (n = 14), (Ullrich 1980; Exo 1983). The following year after the vole population collapsed the average clutch size was only 3.4 eggs (n = 14, p < 0.01, Mann

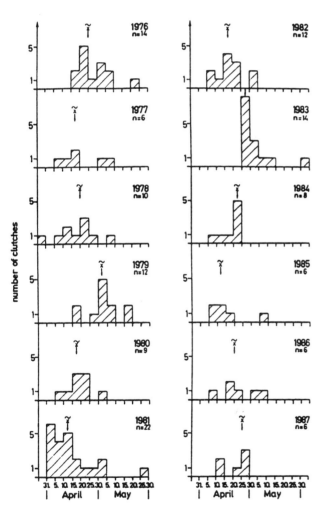

Figure 2. Laying dates of Little Owls in the northern foothills of the Swabian Alb, North-Wuerttemberg 1976–1987 (Ullrich unpubl., replacement clutches excluded). In years with high Common Vole density (e.g. 1981, 1985) egg laying started earlier than in years with low vole density (e.g. 1976, 1979, 1983).

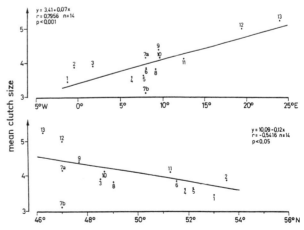

Figure 3. Geographical variation in clutch size of Little Owls (for area numbers and references see Table 3).

Table 3. Breeding success of Little Owls in different European countries.

Location and study period (numbers refer to figure 3)	mean date of first egg	mean clutch size (n)	mean brood size at hatch in successful nests	mean brood size at hatch in nests	mean brood size at fledging in successful nests	mean brood size at fledging in nests	% losses of eggs and nestlings	References
Great Britain								
1. England & Wales 1936–75	28.04.	3.5 (182)			2.4 (241)		50.9	Glue & Scott (1980)
2. Lincolnshire & Nottinghamshire 1950–77		3.9 (86)						Glue & Scott (1980)
France								
3. Eure-et-Loire 1918–50		3.9 (80)						Labitte (1951)
West Germany								
4. Kleve 1974–84		3.6 (96)	3.2 (63)	2.5 (81)	2.8 (61)	1.9 (89)	47.2	Exo (1987)
5. Tecklenburg 1977–85		3.7 (153)						Kimmel pers. comm.
6. Werl 1974–82	22.04.	3.9 (154)				2.0 (186)		Illner pers. comm.
8. Heilbronn 1977–87		3.8 (101)	3.7 (80)	2.9 (101)	3.3 (71)	2.3 (101)	39.2	Furrington pers. comm.
9. Friedrichshafen 1973–88	26.04.	4.4 (265)		3.5 (265)		2.7 (265)	39.7	Knötzsch pers. comm.
10. Göppingen 1969–88	22.04.	4.1 (161)	3.8 (203)	2.8 (203)	3.3 (143)	2.4 (203)		Ullrich pers. comm.
11. East Germany 1960–85		4.1 (126)	3.5 (75)	2.4 (111)	2.8 (110)	1.9 (167)	51.7	Schönn (1986)
7. Switzerland 7a ?		4.2 (30)						Glutz (1962)
7b 1973–80		3.1 (153)	2.8 (122)	2.2 (153)		1.8 (153)	41.6	Juillard (1984)
Czechoslovakia ?		4.8 (37)						Hudec (1983)
12. Hungary 1910–76		5.0 (89)						Nemeth pers. comm.
Rumania								
13. Siebenbürgen 1901–75		5.2 (21)						Kohl pers. comm.

Whitney U-test; Ullrich 1980). Similar results were also found on the Lower Rhine (Exo 1983). In 1977, a vole peak year, average clutch size was 4.2 eggs (n=9), in 1975 and 1976 only 3.4 (n=9) and 3.2 (n=9, $p<0.05$), respectively. On average female Little Owls laid about 1.0–1.2 eggs more in peak vole years than in low vole years. Though we can assume that voles are the main prey species in the pre-laying period it seems that Little Owls are less affected by the population fluctuations of voles than other European owls: for example, Barn Owls *Tyto alba* (Schonfeld *et al.* 1977), Tawny Owls *Strix aluco* (Southern 1970) and Tengmalm's Owls *Aegolius funereus* (Korpimäki 1981). Breeding failure was therefore not associated with the number and availability of rodents. Even in years with exceptionally low vole density the proportion of non-breeders was not unusually high (Ullrich 1980; Exo 1983, but see Illner 1979). Ullrich (1980) assumed that some females 'refrained' from breeding occasionally, but that must not necessarily be caused by prey availability alone. The number of breeding pairs, clutch size and fledging success of Little Owl, is less influenced by vole density than in other vole-eating owl species. This is related to the fact that Little Owls are food 'generalists'; adapting to feed on alternative prey when voles are scarce (Uttendorfer 1939; Glutz & Bauer 1980; Juillard 1984).

Ullrich (1980) found that the clutch size of Little Owls, in the northern foot-hills of the Swabian Alb, declined as the start date of the laying period advanced. As mentioned above, egg laying started earlier in years of high vole density than in low vole years, so the seasonal trend can at least partly be related to the different prey density.

Breeding success

Mean hatching success varied from 2.8–3.8 young per successful clutch to 2.2–3.5 young per clutch respectively, whilst the number of young fledged per successful clutch varied between 2.4–3.3 and 1.8–2.7 per clutch respectively (Table 3). Of all the eggs found, between 39% and 52% failed to produce fledglings and losses during the egg laying and incubation periods were larger than during the nestling period. All values are overall means. Long-term population studies show great differences between years. For example, between 1974 and 1984 on the Lower Rhine, the mean annual reproduction rate ranged from 1.0 to 3.0 fledged young per clutch (Exo 1987); in Switzerland from 1.3 to 2.5 (1973–1980, Juillard 1984) and in an artificial nesthole population around Lake Constance from 1.9 to 3.6 (Figure 4; Knotzsch pers. comm.).

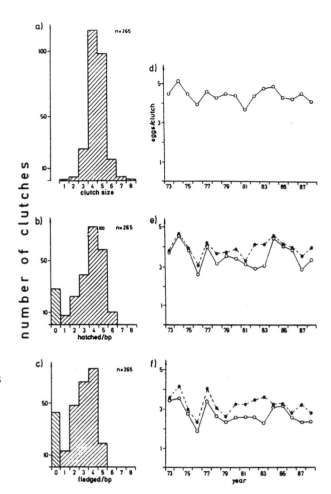

Figure 4. Breeding success of Little Owls in a population near Friedrichshafen/Lake Constance: summary of the breeding data (a-c) and means for each year (d-f) of the study period (Knotzsch unpubl.).

a. clutch size

b. number of hatched young per breeding pair

c. number of fledged young per breeding pair

d. mean clutch size

e. mean number of hatched young per breeding pair (solid line), mean number of hatched young per successful breeding pair (dotted line)

f. mean number of fledged young per breeding pair (solid line), mean number of fledged young per successful breeding pair (dotted line).

The annual breeding success depends on many different factors such as weather conditions, predation rate and age of the breeding birds. The relative importance of each of these factors has yet to be analysed. Exceptionally rainy periods during the nestling period may result in an enhanced nestling mortality (Glutz & Bauer 1980; Finck 1988). Illner (in Glutz & Bauer 1980) found that yearling females produced fewer fledglings, on average, than older females (mean=1.9, n=22 and mean=2.8, n=32 respectively), although older females did not lay more eggs. As mentioned, clutch size was affected by vole abundance, however, to date it is not clear to what extent vole abundance influences the reproduction rate. On the Lower Rhine, Little Owls laid significantly more eggs per clutch and produced significantly more fledglings per successful clutch, in peak vole years than in low vole years. However, the number of young fledged per clutch did not vary significantly (Exo 1987, in prep.). Predation meant that there was no statistically significant relationship between clutch size and reproduction rate. Vole availability had no effect on either the reproductive output or the population regulation. This could at least partly be related to the fact that Little Owls feed their young to a great extent with earthworms (Lumbricidae) and beetles (Coleoptera) (Juillard 1984). There was also no relationship between Little Owl density, on the Lower Rhine, and the reproductive rate.

Mortality

There is less known about the mortality rates of Little Owls than about most other aspects of their ecology. Annual death rates are usually calculated from national ringing schemes or from detailed studies of particular populations. Estimates derived from ringing recoveries cover wide geographical areas and long periods, but they are often biased and therefore must be used with caution (Perdeck 1977; Anderson, Burnham & White 1985; Brownie et al. 1985; Aebischer & Coulson 1987; Exo & Hennes 1980). Studies of particular populations give more accurate information but such studies have only recently been started on Little Owls, and so far very little information is available.

Mortality in the first year of life was estimated at 70.1% (Exo & Hennes 1980), on the basis of national ringing recoveries from West Germany and the Netherlands, while for Little Owls ringed in Switzerland, Glutz & Bauer (1980) calculated a mortality rate, during the first year of life, of 74%. The annual death rates of young birds are influenced to a greater extent by environmental factors, such as climatic conditions, than those of adult birds. Therefore, especially during the first year of life, large variations in mortality may occur from year to year as well as from region to region. For example Knotzsch (1978) found indications of unusually low juvenile mortality rates during the growth phrase of the studied population (Figure 6a). During the breeding season of 1975 he studied 56% (n = 14) of the young raised in 1974 (n = 25) showing that the mortality rate was less than 44%. On the other hand, on the basis of ringing recoveries, juvenile mortality rates between 70% and 74% were estimated.

Most young Little Owls died in the first months after fledging; about 37% of the total of the first year birds died by the end of August, about 57% by the end of October (Exo & Hennes 1980). The high mortality in the first months of life can often be related to the inexperience of the juveniles. In regions with continental, especially snowy, climates a second mortality peak occurs in January or February.

Using recovery data, the annual mortality of a population of Dutch adult Little Owls was calculated at $29.2 \pm 8.5\%$ (n = 19), and that of German Little Owls at $37.9 \pm 5.2\%$ (Exo & Hennes 1980). The higher mortality rate of German than of Dutch Little Owls is probably a result of increased occurrence of winter mortality in Germany; the direct result of different climatic conditions. The annual mortality patterns of adult and first year German Little Owls indicate a mortality peak in January or February. In contrast, Glue (1973), did not find a winter peak in mortality in Great Britain. Populations in the Netherlands seem to take an intermediate position. The higher winter mortality found in German than of British and Dutch Little Owls is the result of the severer winter climates in continental than in oceanic regions (Steinhauser 1970). Extraordinarily high mortality rates in severe, in particular exceptionally snowy, winters have been documented in many papers (Peitzmeier 1952; Poulsen 1957; Piechocki 1960; Dobinson & Richards 1964; Knotzsch 1978; Exo & Hennes 1980; Exo 1983; Schonn 1986). Regional differences in mortality have also been noted in other species, for instance in Tawny Owls (Schifferli 1957; Olssen 1958). The high winter mortality can at least partly explain the northern and the altitudinal limits of the range.

A second peak in adult mortality occurs during the fledging period of the young owls, in June/July. This mortality peak is associated with the stress of rearing the young and the beginning of the moult, which usually starts around the time of fledging. Both result in a higher energy expenditure. Although we can assume that both prey density and prey availability are high during this time of the year, the stress of rearing the young and moult causes an energetic bottleneck (Exo 1988). Whereas the winter mortality peaks occur only as a result of exceptionally high mortality rates in very severe winters, the summer peak in mortality occurs each year.

As previously indicated, clutch size increases from west to east and from north to south in Europe (Figure 3), but whether the mortality rate decreases in a corresponding way is unknown. Furthermore, it is not known whether the mortality of adult Little Owls is constant with age (Exo & Hennes 1980). A third aim for future studies is to determine separate mortality rates for sexes. Knotzsch (1988) found first indications that the annual death rates of females may be higher than of males.

Mortality vs reproduction

The question arises as to whether reproduction is able to compensate for mortality or if the population decrease over wide areas of central Europe is due to lower reproduction, as Juillard (1989) suggested. In stable populations, reproduction balances mortality. Knowing the annual death rates of juvenile and adult Little Owls, we can calculate the reproduction rate which is theoretically necessary to compensate for mortality (Henny et al. 1970). To balance the demonstrated mortality rates of m(ad) = 35% and m(juv) = 70% each pair of Little Owls must produce 2.34 fledglings each year (Exo & Hennes 1980). Assuming that approximately 40% of the eggs found failed to produce fledglings, each pair must lay 3.9 eggs per year (in both calculations failure to breed was not taken into account). Table 3 shows that the calculated required fledging success of 2.35 fledglings per nest is higher than the observed fledging success in some European populations (e.g. Switzerland, East Germany). However, it cannot be concluded that the observed population decrease is exclusively due to poor reproduction, as differences between calculated and observed fledging success can be partly caused by geographical variations in mortality. Furthermore, the calculated values are averages from many decades and the ringing recoveries are often biased in some way. Mortality estimates derived from recoveries can over- or under-estimate mortality. Comparatively small miscalculations of the mortality rates results in a large over- or under-estimation of the reproduction rate. For example, if the annual death-rates of juvenile and adult Little Owls are only 5% lower (65% and 30% respectively) than calculated from the recoveries, each pair must produce only 1.7 fledged young per year to compensate mortality, compared with 2.35 assuming 70% and 35% respectively. The necessary fledging success of 1.7 young was obtained in all studied populations (Table 3).

At the present time, there are data for only one study area which allow comparison of a theoretically calculated population trend with an observed trend. The area, the Lower Rhine, is characterised by a comparatively high Little Owl density but extremely low reproduction: between 1974 and 1984 mean density was about 1.7 pairs/km^2, mean reproduction rate about 1.9 fledged young/clutch (Table 3, Exo 1987, in prep.). Although the reproduction rate was lower than in most other European populations, Little Owl density was obviously still above that

needed to maintain a stable population. Winter losses of about 22% were balanced in the following 2–3 years. For the Lower Rhine it is presumed that in years with high population density or with high reproductive output, above the long-term average, the population produced additional birds that migrated to other areas (Exo 1983, 1987, in prep.). In order to answer the question of whether the noted population decrease over much of Europe was due to poor reproduction or enhanced mortality, detailed long-term population studies from different geographical and climatic regions, especially analysis of the annual mortality rates of adult and juvenile birds, are required.

The role of nest-site availability in limiting breeding density

In Central Europe, Little Owls prefer sparsely wooded lowland areas with year-round short vegetation provided these areas are well covered with old pollarded trees, orchards and farm buildings affording nest-holes, day-roosts and hunting look-out posts, such as fence posts. Some recent studies concerned with habitat structure and nest site ecology give evidence that population density is limited by breeding sites and/or land utilisation (Visser 1977; Exo 1983; Loske 1986; O'Connor & Shrubb 1986; Epple & Holzinger 1987). Figure 5 gives a typical example; the figure shows a positive correlation between the number of Little Owl territories and the number of possible breeding tree sites for the Lower Rhine area. A detailed study of nest site ecology confirmed that density and distribution patterns were determined by the occurrence of suitable breeding trees. 85% of the breeding pairs inhabited cavities in trees, mainly in pollarded willows, though many suitable nesting facilities in buildings were available (Exo 1981). Similar relationships between the number of Little Owl territories and the number of possible breeding tree sites were noted in some other areas, for example in the Maas-Waal lowlands near Nijmegen (Visser 1977) and in Middle Westphalia (Loske 1986).

Furthermore, Exo (1983) and Loske (1986) found that Little Owl density increased significantly in parallel with an increase in area of grassland. This correlation was at least partly related to a spurious positive correlation between the number of suitable breeding trees and the proportion of grassland. Although tree density was related to the occurrence of grassland it seems to be very probable that the Little Owl density was affected by the occurrence of grassland, year-round short vegetation and of suitable breeding trees. The other part of the correlation between the proportion of grassland and

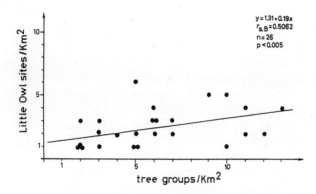

Figure 5. Little Owl density in relation to the number of potentially suitable 'natural' nest sites (especially pollarded tress and old fruit-trees) on 26 plots of 1 km each (for details see Exo 1983).

population density can be explained by the preferred hunting technique. Radio-tracking and direct observations with a light intensifier showed that the Little Owl is a typical ground-hunter. Most hunting is done by running or hopping over the ground; they hunt from perches to a much lesser extent (Exo 1987, in prep.). Hence it follows that Little Owls, in Central Europe, are dependent on areas with short vegetation throughout the year, especially grassland habitats. Furthermore, grassland habitats in general have higher earthworm and insect densities than other areas which are used as arable land (Tischler 1965, 1980). Investigations of radio-tagged Little Owls confirmed this, since hunting birds used grassland habitats much more than expected from the area available (Exo 1987; Finck 1988).

The clearest evidence for limitation of population density by the number of available nest sites was illustrated by investigations which used artificial nest boxes in substitution for the loss of natural holes (Figure 6; Knotzsch 1978, 1988; Epple & Holzinger 1987). As Figure 6 illustrates, population density increased until the carrying capacity of the particular area was reached. In the first years after the nest boxes were provided, population density increased rapidly. However after 3–5 years when the population neared the carrying capacity of the habitat, the growth rate diminished once again from year to year. These studies showed firstly that the population density was limited by suitable nesting places, and secondly that there were further limiting factors. Although Little Owls are dependent on grassland habitats, since they are ground-hunters, the availability of nesting places in Central Europe is at present the limiting factor. When the limiting factor is eliminated, density increases until the population reaches a new limit. Knotzsch (1988) suggested that the population at Lake Constance (Figure 6a), after a

sufficient number of nest boxes had been provided, was limited by climatic conditions.

Density-dependent population regulation

The aforementioned investigations support the conclusion that number and distribution of nest-sites are the 'ultimate' limiting factors determining population density across the breeding range of Little Owls in Central Europe. The present knowledge on the 'proximate' causes of population regulation are discussed in the next part of the paper.

The growth rate of populations supported by nest-boxes diminished from year to year as population density increased. This, as well as the stability in numbers after reaching the carrying capacity, may give a first indication of density-dependent population regulation. A standard method to examine population regulation is the 'key-factor

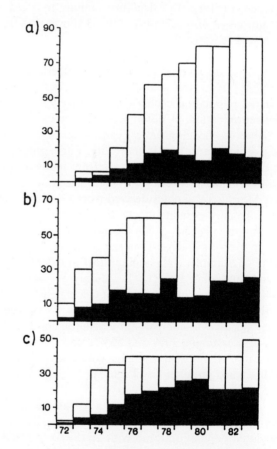

Figure 6. Little Owl population trends in comparison to the number of artificial nest-sites for three study areas:

a.   Friedrichshafen/Lake Constance (Knotzsch)

b.   Northern foothills of the Swabian Alb (Ullrich)

c.   district of Heilbronn (Furrington)
     Black columns = number of breeding pairs, white columns – number of artificial nests
     (from Epple & Holzinger 1987).

analysis' developed by Varley & Gradwell (1960). The key-factor analysis allows quantification of the effects of variations in natality and mortality and recognition that these factors operate in a density-dependent manner. Until now this method has been used to analyse population data of two different Little Owl populations: an artificial nest hole population at Lake Constance (Knotzsch 1988) and the Lower Rhine population, a natural population which was not influenced by the provision of nest-boxes (Exo 1983, 1987, in prep.).

At both sites 'overwinter loss' (losses occurring outside the breeding season) was clearly the key-factor which influenced changes in annual breeding density ('overwinter loss' includes mortality together with immigration and emigration). Overwinter loss was related most closely to the total annual loss, whereas losses during the breeding season contributed less to the total annual mortality. On the Lower Rhine, overwinter loss explained about 56% of the total annual mortality. Furthermore, losses during the breeding season were inversely correlated with subsequent overwinter loss, suggesting that a decline in overwinter losses compensated losses during the previous breeding season.

The various losses during breeding explained 44% of the annual losses: 45% occurring during the incubation period, 35% during the nestling period and only about 20% related to the annual variation of clutch size. Clutch size varied in accordance with vole abundance in the pre-laying period but the reproduction rate was not related to vole abundance. Even if it is assumed that beeding failure is higher in low vole years than in peak vole years, the annual variation of the numbers of eggs laid by the whole population will probably not affect population changes (Exo 1987, in prep.).

The primary importance of the overwinter losses is evident when the losses occurring outside the breeding season are plotted against the losses throughout the previous breeding period. On the Lower Rhine, losses during breeding were inversely correlated with subsequent overwinter losses. When losses were high outside the breeding season, losses up to the fledging stage had been small in the previous breeding season, and vice versa. So, reduced overwinter loss compensated losses during the previous breeding season. The same result was obtained when the annual overwinter losses were plotted against the number of young fledged. As the number of young fledged increased, overwinter loss increased too. Overwinter loss was significantly correlated with the summer density. The exponential relationship between overwinter loss and density

indicates that the population was regulated in a density-dependent way when the carrying capacity was reached.

Overwinter loss primarily refers to the high mortality of the juveniles; about 50–75% of the juveniles died in their first year. On the Lower Rhine an increase in winter mortality occurred only in very severe winters. The area was predominantly influenced by maritime air masses. In contrast to other regions (Knotzsch 1988) on the Lower Rhine, winter mortality affected population changes only in exceptional circumstances and in general population density was not limited by climatic conditions. It can be assumed that in years with high population density in the late summer (due to high breeding density and/or high reproduction rate) the population produced additional birds that were excluded from the study area by the territorial behaviour of the adults. Some juveniles therefore emigrated after reaching independence whilst others settled in less suitable habitats within the study area.

Territorial behaviour acts as a proximate regulating factor, whilst density was ultimately limited by suitable nest-sites. Radio-tracking studies and systematic catches using mist-nets showed that some juvenile Little Owls occupied day-roosts in their first autumn/winter which were unsuitable as an established breeding territory due to a lack of suitable nesting holes and hunting areas. From these areas, juveniles intruded regularly into suitable but occupied territories (Exo 1987, in prep.). These observations support the assumption that on the Lower Rhine the breeding density was limited by the availability of natural resources, especially nest-sites and hunting grounds, and not by natality or mortality. For other regions, however, we have to assume that mortality, especially winter mortality, limits population density (Knotzsch 1988).

Acknowledgements

I want to express my deep gratitude to Prof. Dr D. Neumann for giving me the opportunity to carry out fieldwork on Little Owls for more than ten years and for his stimulating coaching during that time. I am very grateful to G. Knotzsch and Dr B. Ullrich who made their unpublished data freely available for incorporation in this review. Mr A. Streich kindly drew the figures and Mr K. Wilson improved the English.

References

Aebischer, N.J. & Coulson, J.C. 1987. Lower juvenile survival rates estimated from ringing returns: reality or artefact. *Ibis, 129*: 116–117.

AG Zum Schutz Bedrohter Eulen Nrw 1978. Steinkauz-Verbreitung. *Informationsblatt Nr. 7*: 5–6.

Anderson, D.R., Burnham, K.P. & White, G.C. 1985. Problems in estimating age-specific survival rates from recovery data of birds ringed as young. *Journal of Animal Ecology, 54*: 89–98.

Brownie, D., Anderson, D.R., Burham, K.P. & Robson, D.S. 1985. *Statistical inference from band recovery data – a handbook*. U.S. Fish and Wildlife Service, Resource Publication No. 156, 2nd ed., Washington D.C.

Cramp, S. & Simmons, K.E.L. eds. 1985. *Handbook of the birds of Europe, the Middle East and North Africa*. Vol. 4, Oxford, Oxford University Press.

Davies, N.B. 1978. Ecological questions about territorial behaviour. *In*: *Behavioural ecology – an evolutionary approach*, ed. by J.E. Krebs & N.B. Davies, 317–350. Oxford, Blackwell Scientific Publications.

Dobinson, H.M. & Richards, A.J. 1964. The effects of the severe winter of 1962/63 on birds in Britain. *British Birds, 57*: 373–434.

Epple, W. & Holzinger, J. 1987. Steinkauz. *In*: *Die Vogel Baden-Wurttembergs*, ed. by J. Holzinger. Vol. 1.2: 1085–1095, Karlsruhe.

Exo, K.M. 1981. Zur Nistokologie des Steinkauzes (*Athene noctua*). *Vogelwelt, 102*: 161–180.

Exo, K.M. 1983. Habitat, Siedlungsdichte und Brutbiologie einer niederrheinischen Steinkauzpopulation (*Athene noctua*). *Okol. Vogel, 5*: 1–40.

Exo, K.M. 1987. Das Territorialverhalten des Steinkauzes (*Athene noctua*) – eine verhaltensokologische Studie mit Hilfe der Telemetrie. Disseration of the University of Koln.

Exo, K.M. 1988. Jahreszeitliche okologische Anpassungen des Steinkauzes (*Athene noctua*). *Journal of Ornithology, 129*: 393–415.

Exo, K.M. & Hennes, R. 1978. Empfehlungen zur Methodik von Siedlungsdichte-Untersuchungen am Steinkauz (*Athene noctua*). *Vogelwelt, 99*: 137–141.

Exo, K. M. & Hennes, R. 1980. Beitrag zur Populationsokologie des Steinkauzes (*Athene noctua*) – eine Analyse deutscher und niederlandischer Ringfunde. *Vogelwarte, 30*: 162–179.

Finck, P. 1988. Variabilitat des Territorialverhaltens beim Steinkauz (*Athene noctua*). Dissertation of University of Koln.

Genot, J.-C. 1988. Ecologie et protection de la Chouette cheveche (*Athene noctua* Scop.). – Duexieme partie: Habitat, reproduction, regime alimentaire. Parc Naturel Region des Vosges du Nord.

Glue, D.E. 1973. Seasonal mortality in four small birds of prey. *Ornis Scandinavica, 4*: 97–102.

Glue, D.E. & Scott, D. 1980. Breeding biology of the Little Owl. *British Birds, 73*: 167–180.

Glutz von Blotzheim, U.N., ed. 1962. *Die Brutvogel der Schweiz*. Arau.

Glutz von Blotzheim, U.N. & Bauer, K.M. 1980. *Handbuch der Vogel Mitteleuropas. Bd. 9 (Columbiformes – Piciformes)*. Akademie Verlagsgesellschaft Wiesbaden.

Haverschmidt, F. 1946. Observations on the breeding habitats of the Little Owl. *Ardea, 34*: 214–246.

Hegger, H.L. 1977. Steinkauz, Waldkauz und Waldohreule als Brutvogel im Kempener Land. *Heimatbuch des Kreises Viersen, 1977*: 58–63.

Henny, C.J., Overton, W.S. & Wright, H.M. 1970. Determining parameters for populations by using structural models. *Journal of Wildlife Management, 34*: 690–703.

Hudec, K., ed. 1983. *Fauna CSSR, Ptaci-Aves. Vol. 3/1*. Prague.

Huxley, J.S. 1934. A natural experiment on the territorial instinct. *British Birds, 27*: 270–277.

Illner, H. 1978. Eulenbestandsaufnahme auf dem MTB Werl von 1974–1978, ed. AG zum Schutz bedrohter Eulen. *Informationsblatt, 9*: 5–6.

Illner, H. 1988. Langristiger Ruckgang von Schleiereule (*Tyto alba*), Waldohreule (*Asio otus*), Steinkauz (*Athene noctua*) und Waldkauz (*Strix aluco*) in der Agrarlandschaft Mittelwestfalens. *Vogelwelt, 109*: 145–151.

Juillard, M. 1979. La croisance des jeunes Chouette chevches (*Athene noctua*), pendant leur sejour au nid. *Nos Oiseaux, 35*: 113–124.

Juillard, M. 1984. La chouette cheveche. *Nos Oiseaux, 1984, Sonderheft*.

Juillard, M. 1989. The decline of the Little Owl (*Athene noctua*) in Switzerland. *In*: *Raptors in the modern world*, ed. by B.U. Meyburg & R.D. Chancellor, 435–439. Berlin, WWGBP.

Kampfer, A. & Lederer, W. 1988. Dismigration des Steinkauzes (*Athene noctua*) in Mittelwestfalen. *Vogelwelt, 109*: 155–164.

Kesteloot, E.J.J. 1976. Present situation of birds of prey in Belgium. *ICBP World conference on birds of prey, Vienna 1975*: 85–87.

Kempf, C. 1973. La rapaces nocturnes d'Alsace. *Alauda, 41*: 413–418.

Klomp, H. 1970. The determination of clutch-size in birds: a review. *Ardea, 58*: 1–124.

Knotsche, G. 1978. Ansiedlungsversuche und Notizen zur Biologie des Steinkauzes (*Athene noctua*). *Vogelwelt, 99*: 41–54.

Knotsche, G. 1988. Bestandsentwicklung einer Nistkasten-Population des Steinkauzes (*Athene noctua*) am Bodensee. *Vogelwelt, 109*: 164–171.

Korpimäki, E. 1981. On the ecology and biology of Tengmalm's Owl (*Aegolius funereus*) in Southern Ostrobothnia and Suomenselka, Western Finland. *Acta Univ. Oul., A 118, Biol. 13*: 1–84.

Kostrzewa, R. 1988. Die Dichte des Turmfalken (*Falco tinnunculus*) in Europa. Ubersicht und kritische Betrachtung. *Vogelwarte, 34*: 216–224.

Labitte, A. 1951. Notes biologiques sur la Chouette cheveche. *L'Oiseau, 21*: 120–129.

Lack, D. 1947. The significance of clutch-size. *Ibis, 89*: 302–352.

Loske, K.H. 1986. Zum Habitat des Steinkauzes (*Athene noctua*) in der Bundesrepublik Deutschland. *Vogelwelt, 107*: 81–101.

Maynard Smith, J. 1974. *Models in ecology*. Cambridge, Cambridge University Press .

Mildenberger, H. 1984. *Die Vogel des Rheinlandes*. Vol. 2. Dusseldorf.

Murray, G.A. 1976. Geographic variation in the clutch sizes of seven owl species. *Auk, 93*: 602–613.

Newton, I. 1979. *Population ecology of raptors*. Calton, Poyser.

O'Connor, J.R. & Shrubb, M. 1986. *Farming and Birds*. Cambridge, Cambridge University Press.

Olsson, V. 1958. Dispersal, migration, longevity and death causes of *Strix aluco, Buteo buteo, Ardea cinerea* and *Larus argentatus*. *Actua Vertebrata, 1*: 85–189.

Peitzmeier, J. 1952. Langsamer Ausgleich der Winterverluste beim Steinkauz. *Vogelwelt, 73*: 136.

Perdeck, A.C. 1977. The analysis of ringing data: pitfalls and prospects. *Vogelwarte, 29*: 33–44, Sonderheft.

Petzold, H. & Raus, T. 1973. Steinkauz (*Athene noctua*) – Bestandsaufnahmen in Mittelwestfalen. *Anthus, 10*: 25–38.

Piechocki, R. 1960. Uber die Winterverluste der Schleiereule (*Tyto alba*). *Vogelwarte, 20*: 121–124.

Poulsen, C.M. 1957. Massedodsfald blandt Kirkengler (*Athene noctua* [Scop.]). *Dansk Ornitologisk Forenings Tidsskniff, 51*: 30–31.

Schifferli, A. 1957. Alter und Sterblichkeit bei Waldkauz (*Strix aluco*) und Schleiereule (*Tyto alba*). *Der Ornithologische Beobachter, 54*: 50–56.

Schonfeld, M., Girbig, G. & Sturm, H. 1977. Beitrage zur Populationsdynamik der Schleiereule, *Tyto alba. Hercynia N.F., 14*: 303–351.

Schonn, S. 1986. Zu Status, Biologie, Okologie und Schutz des Steinkauzes (*Athene noctua*) in der DDR. *Acta Ornithoecol, 1*: 103–133.

Schonn, S., Scherzinger, W., Exo, K.M. & Ille, R. 1990. *Der Steinkauz*. Neue Brehm-Bucherei.

Sharrock, J.T.R. 1976. *The atlas of breeding birds in Britain and Ireland*. Berkhamsted, BTO/IWC.

Southern, H.N. 1970. The natural control of a population of Tawny Owls (*Strix aluco*). *Journal of Zoology, London, 162*: 197–285.

SOVON 1987. *Atlas van de Nederlandse Vogels*. Arnhem, SOVON.

Steinhauser, F. ed. 1970. *Climatic atlas of Europe. I. Maps of mean temperature and precipitation*. Geneva, WMO Unesco, Cartographia.

Tischler, W. 1965. *Agrarokologie*. Jena.

Tischler, W. 1980. *Biologie der Kulturlandschaft*. Stuttgart, Gustav Fischer Verlag.

Tompa, E.S. 1962. Territorial behaviour: the main controlling factor of a local song sparrow population. Reprint in: Stokes, A.W., ed. 1974. *Territory*. Stroudsberg, Dowden, Hutchison & Ross.

Ullrich, B. 1973. Beobachtungen zur Biologie des Steinkauzes (*Athene noctua*). *Anz. orn. Ges. Bayern, 12*: 163–175.

Ullrich, B. 1980. Zur Populationsdynamik des Steinkauzes (*Athene noctua*). *Vogelwarte, 30*: 179–198.

Uttendorfer, O. 1939. *Die Ernahrung der deutschen Tagraubvogel und Eulen (und ihre Bedeutung in der heimischen Natur)*. Neumann-Neudamm.

Varley, G.C. & Gradwell, G.R. 1960. Key factors in population studies. *Journal of Animal Ecology, 29*: 399–401.

Village, A. 1984. Problems in estimating Kestrel breeding density. *Bird Study, 31*: 121–125.

Visser, D. 1977. De Steenuil in het Rijk van Nijmegen. *De Mourik, 3*: 13–27.

Voous, K.H. 1960. *Die Vogelwelt Europas und ihre Verbreitung*. Hamburg, Parey.

Witt, K. 1986. Bestandwerfassung einiger ausgewahlter Vogelarten (1982–1984) in der Bundesrepublik Deutschland. *Vogelwelt, 107*: 231–239.

Yeatman, L. 1976. *Atlas des oiseaux nicheurs de France de 1970 a 1975*. Paris.

Ziesemer, F. 1981. Zur Verbreitung und Siedlungsdichte des Steinkauzes (*Athene noctua*) in Schleswig-Holstein. *Zool. Anz., Jena, 207*: 323–334.

Productivity and density of Tawny Owls *Strix aluco* in relation to the structure of a spruce forest in Britain

S.J. Petty & A.J. Peace

Petty, S.J. & Peace, A.J. 1992. Productivity and density of Tawny Owls *Strix aluco* in relation to the structure of a spruce forest in Britain. *In: The ecology and conservation of European owls*, ed. by C.A. Galbraith, I.R.Taylor and S. Percival, 76-83. Peterborough, Joint Nature Conservation Committee. (UK Nature Conservation, No. 5.)

1. A population of Tawny Owls *Strix aluco* was monitored from 1981 until 1987 in approximately 90 km^2 of short-rotation spruce forest in northern England, where the owls fed largely on Field Voles *Microtus agrestis*.

2. Vole populations showed a cyclic trend with a peak every third year. This led to great annual variations in the number of young owls reared.

3. The productivity of individual owl territories varied greatly over the seven year period, but was not density-dependent.

4. A multiple regression model incorporating three habitat categories explained 68% of the variability in productivity and emphasised the importance of habitat heterogeneity.

5. The density of owl territories increased with the spatial diversity of the habitat.

S.J. Petty, Forestry Commission, Wildlife & Conservation Research Branch, Ardentinny, Dunoon, Argyll PA23 8TS.

A.J. Peace, Forestry Commission, Forestry Research Station, Alice Holt Lodge, Wrecclesham, Farnham, Surrey GU10 4LH.

Introduction

In the last 60 years, large areas of the British uplands have been planted with exotic conifers; spruce *Picea* spp. have been increasingly used (Anon 1984). Most crops are grown on short rotations (45–55 years) and many of the earlier plantings are now being felled and replanted (restocked). A number of studies have looked at the effects of the afforestation of open ground on raptors (Newton 1983; Ratcliffe & Petty 1986; Petty 1988) but little is known about how birds of prey react to restocking.

The Tawny Owl *Strix aluco* is a territorial, nocturnal predator of small mammals (Southern 1970; Cramp 1985), and often colonises spruce forests by the end of the first rotation. Hirons (1985) studied the size of Tawny Owl territories in a spruce forest in south Scotland, and showed that they were larger than in mixed farmland and broadleaved woodland in southern England. He also noted a lack of data about the areas used by the owls for hunting, and about the distribution of their prey in conifer forests. Petty (1987a) showed that Tawny Owls in a spruce forest in northern England fed mainly on the Field Vole *Microtus agrestis*, and that newly restocked sites with grassy vegetation provided a major vole habitat. The number of chicks produced annually by, between 47–50 pairs of owls, ranged from seven in a year with low vole numbers to 142 in a year when vole populations peaked.

In addition to this annual variation in productivity, there was great variation between territories in the number of chicks produced. This paper investigates whether differences in spatial arrangement of habitat types in a spruce forest can account for variations in the density and productivity of Tawny Owl territories over a seven year period. Throughout this period the configuration of habitats in the forest altered dramatically. An area of 1000 ha was felled and flooded for a reservoir, and each year large blocks of forest were felled and replanted creating a patchwork of multi-aged stands.

Methods

Study area

The study area in Kielder Forest, northern England (55° 3'N, 2° 3'W), was part of a more extensive tract of forest which has been planted on the border between England and Scotland (Figure 1). The study area measures about 90 km^2 and comprises largely plantations of Sitka Spruce *Picea sitchensis* and Norway Spruce *Picea abies* grown on rotations of 45–55 years. The forest was entirely man-made, afforestation having commenced in 1933, while clear

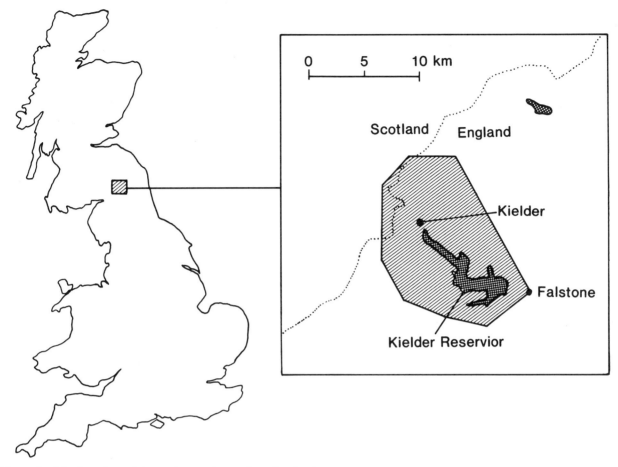

Figure 1. The location of the study area in northern England.

felling and restocking with mainly Sitka Spruce started in 1968. By 1987 there was an extensive patchwork of first and second generation tree crops, particularly at lower elevations. In 1981–1982, 1000 ha below 185 m elevation was flooded to form Kielder Water reservoir.

The Tawny Owl population

Most of the Tawny Owls bred in nestboxes, many of which had been erected prior to 1981, although some boxes were added in subsequent years (Petty 1987b). Between 80% and 100% of the females were caught each time they bred by netting them at the nest-site after the chicks became five days old.

Many of the captured owls were of known identity, having been ringed in the past as breeding females or chicks. Over the years it was possible to determine groups of nestboxes and natural sites that had been used by individual females. There were 58 such 'nesting territories', which were regularly spaced along all the major valleys in the study area (Figure 2). In the present analysis, 30 of these nesting territories were selected, where the results of each

breeding attempt during 1981–1987 were known (Figure 2). Only once in these 210 territory years was a territory unoccupied. The remaining 28 territories were excluded from this analysis as breeding data for some of the earlier years were not collected. There were no predators in the study area which could remove Tawny Owl chicks or eggs from nestboxes.

In the studies of Southern (1970) and Hirons (1985), the extent of the Tawny Owl territories was determined from the position of calling birds. I used a different method, based on the nearest neighbour distance between owl pairs, which was not an accurate measure of the areas used by owls but an estimate of the habitats available around each nesting territory.

The Field Vole population

In 1984–1987 a trapping method was used three times each year to determine fluctuations in the Field Vole population on one restocked site (Petty 1987a). A quicker method of estimating vole abundance was also developed which used vole signs (faeces and heaps of grass clippings in vole runs).

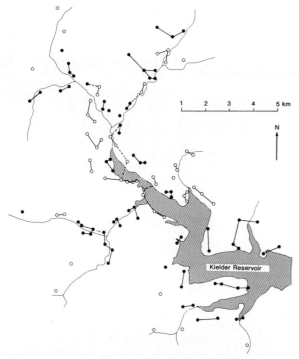

Figure 2. The location of 58 Tawny Owl nesting territories in Kielder Forest. Circles joined by a solid line show alternative nest sites in the same territory, circles joined by a broken line indicate where adjacent pairs sometimes use the same nest site. The filled circles show the 30 nesting territories used in the present analysis.

This method was used in an additional 13 sites in the study area during 1984–1987 (Petty unpublished data). For the years 1981 to 1983, vole abundance was subjectively scored using vole signs on a scale of one (low) to 10 (high). These methods were adequate to classify each of the seven years into three categories based on vole abundance, these were:

High   Population relatively high throughout the winter and increasing through the year to a peak in the autumn/winter.

Declining   High population in the winter, but declining by the spring, then remaining low throughout the summer and autumn.

Low   Low population through the winter and spring increasing by the autumn.

Habitat classification and measurement

In each of the 30 nesting territories, 13 habitat variables were measured (Table 1). The centre of each owl's nesting territory was fixed at the nest site when only one was used, or at the mid-point between nests when more than one had been used. The nearest neighbour distance (NND) was the linear measurement to the centre of the nearest neighbouring owl's territory (Newton *et al.* 1977).

The size of each nesting territory (ARA), was calculated from the area of a circle with a radius of half the NND. The different habitat categories within this circle were then measured to the nearest 0.5 ha from Forestry Commission Stock Maps (scale 1 : 10,000); for some of the analysis these values were converted to proportions of the nesting territory area (ARA). Habitat edge (used in variates EDG and DIV) was the linear length of distinct ecotones between any of the last eight habitat types in Table 1.

Results

Annual variation in the productivity of Tawny Owls

During the seven year period, vole populations were 'high' in three years (1981, 1984 and 1987), 'declining' in two years (1982 and 1985), and 'low' in two years (1983 and 1986) (Figure 3). This cyclic trend in the owls' food supply peaked every third year and corresponded to significant differences in the number of chicks fledged (Table 2). In poor food years, a high proportion of the female owls failed to lay eggs, while in good food years as many as five chicks were occasionally produced by individual pairs (Petty 1987a).

Table 1. Habitat variate acronyms for measurements taken in 30 Tawny Owl territories. The areas of the last 8 variates are measured within ARA. (See Ratcliffe & Petty 1986 for definitions of the forest growth stages.)

Variate name	Description of habitat measurement
NND	Nearest neighbour distance (km) – see text.
ALT	Altitude (m.a.s.l.) at the centre of the nesting territory.
ARA	Area (ha) enclosed by a circle with a radius of NND/2 drawn from the centre of a nesting territory.
EDG	Total length (m) of habitat edges within ARA.
DIV	The length of habitat edges per ha (EDG/ARA).
MAT	Ha planted prior to 1935. Extended rotation broadleaf and conifer forest with well developed ground vegetation.
PFL	Ha planted 1935–1949. Pre-felling conifer forest with little ground vegetation.
THK	Ha planted 1950–1969. Thicket conifer forest with no ground vegetation.
PTK	Ha planted 1970–1980. Pre-thicket conifers with dense ground vegetation
EST	Ha planted 1981–1985. Recently established conifers with developing ground vegetation.
GRZ	Ha of grassland grazed by domestic stock.
UNG	Ha of grassland ungrazed by domestic stock.
WAT	Ha of water.

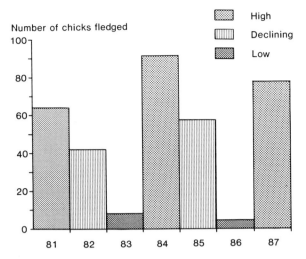

Figure 3. Annual variation in the number of Tawny Owl chicks fledged from 30 territories in each of seven years. High, declining and low refer to the relative abundance of Field Voles.

Table 2. Variation in Tawny Owl productivity in relation to Field Vole abundance. The difference between the mean number of chicks fledged in high compared to declining vole years is very highly significant (t = 4.40, df = 148, p < 0.001).

Years	Field Vole population	Mean number of chicks fledged per territory S.E. ± (n)
1981, 1984, 1987	High	2.58 ± 0.16 (90)
1982, 1985	Declining	1.65 ± 0.14 (60)
1983, 1986	Low	0.20 ± 0.07 (60)

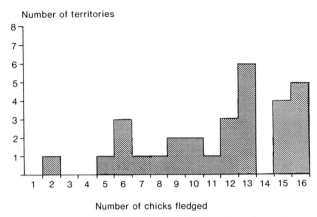

Figure 4. Long-term variation in the total number of Tawny Owl chicks fledged from each of 30 territories over a seven-year period, the median was 12 chicks.

Table 3. Summary statistics for the habitat variates. Descriptions of variate acronyms are given in Table 1.

Variate	Mean	S.E.	Range
NND	0.89 km	0.06	0.32–1.58
ALT	220.67 m	5.07	190.00–320.00
ARA	69.87 ha	9.04	8.00–196.00
EDG	3054.67 m	256.47	780.00–6020.00
DIV	58.37 m	5.68	19.00–146.00
MAT	0.63 ha	0.26	0.00–5.50
PFL	23.43 ha	4.76	0.00–101.00
THK	19.95 ha	7.40	0.00–159.00
PTK	5.65 ha	1.75	0.00–43.00
EST	2.37 ha	0.93	0.00–25.00
GRZ	3.90 ha	1.23	0.00–32.00
UNG	4.37 ha	1.01	0.00–22.00
WAT	9.57 ha	3.00	0.00–64.00

Long-term variations in the productivity of individual Tawny Owl territories

One way of measuring the quality of individual Tawny Owl territories, is to look at the total number of chicks produced over a number of years. This overcomes annual variations in chick production due to fluctuating food supplies (Table 2). Over the seven-year period, the productivity of the 30 territories varied greatly, with the poorest fledging only two chicks and the best producing 16 (Figure 4).

Relationship between habitat composition and Tawny Owl productivity

A summary of the habitat statistics is given in Table 3. Territories calculated from nearest neighbour distances were on average 70 ha, which were larger than Hirons (1985) found in another spruce forest in south Scotland by territory mapping. The average territory comprised 63% closed canopy forest (MAT, PFL and THK), 11% early (open) forest growth stages (PTK and EST), 11% grassland (GRZ and

UNG) and 14% water (WAT). Absolute and proportional habitat measures were initially used in the analysis, but as there was little difference between the two, all the reported analyses used absolute measures.

Analysis of the difference in Tawny Owl productivity could not be satisfactorily explained by any one single habitat variable. The best of the individual habitat models, the linear regression of numbers of fledged owls upon area of water, was significant (p < 0.01) but only explained 28% of the variation. A more complex model was required. The relationship between number of fledged chicks and the area of pre-felling conifer is curvilinear (Figure 5), and suggests a possible optimal amount of pre-felling conifer (PFL) to be contained within each owls territory. Other habitat categories showed a similar curvilinear relationship with the number of chicks reared. If habitat heterogeneity is an important factor in successful Tawny Owl reproduction, then one would have expected the fit of a bivariate linear regression model to be poor, as was the case, and a

79

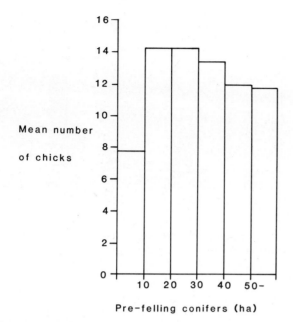

Figure 5. Relationship between the mean number of chicks fledged from 30 territories over a seven-year period, and the area of pre-felling conifers (PFL) in an owls territory. A quadratic model fitted to a scatter plot of this data is highly significant ($p < 0.01$) while a linear model is not significant ($p > 0.05$).

b. The number of owls reared showed a quadratic relationship with the amount of pre-felling conifer (Figure 5).

c. The number of fledged owls had a quadratic relationship with the amount of ungrazed grassland, between 8–9 ha produced the most chicks.

The effect of food supply on owl productivity was also investigated by looking at the relationship between the number of chicks fledged within each territory for the high, declining and low vole years (Table 5). If certain territories contained an optimum habitat composition, or if certain owls were more prolific than others, then the better territories should have constantly produced more chicks compared to the poorer territories, independent of vole abundance. This was not the case (Table 5). The best territories for rearing chicks in high vole years were also among the most successful in declining vole years but in low vole years the most successful territories were uncorrelated with those of the high and declining vole years.

The most successful territories in each of the three vole abundance classes exhibited significant differences in habitat composition (Table 6). In high and declining vole years the most successful territories contained relatively larger amounts of pre-felling conifer (PFL), ungrazed grassland (UNG) and pre-thicket conifers (PTK). In other words most chicks were produced from territories with some good vole habitat (UNG and PTK) opposed

multiple regression model to be more appropriate. Square transformations of the habitat variables were added to the explanatory set of variables and a stepwise multiple regression analysis produced the following equation using three variables:

Total chicks $= 8.44 - 0.00159$ WAT $+ 0.238$ PFL

$- 0.00253$ PFL$^2 + 0.665$ UNG $- 0.0386$ UNG^2

In this equation all three variables are statistically significant ($p < 0.05$) and 68% of the variation is accounted for (Table 4). The model can be interpreted as follows:

a. The numbers of fledged owls decreased with the amount of water in the territory.

Table 5. Spearman Rank correlation coefficients showing the relationship between the number of young Tawny Owls fledged in high, declining and low vole years within each of 30 territories. ($* = p < 0.05$).

	High	Declining	Low
High	1.00		
Declining	0.46*	1.00	
Low	−0.27	−0.15	1.00

Table 4. A stepwise multiple regression model using 3 habitat variable explained 68% of the variation in the number of Tawny Owl chicks reared from 30 nesting territories over a 7-year period.

		S.E.	t	Cumulative variation accounted for
Constant	8.44	0.94	8.97	
WAT (squared)	−0.00159	0.000514	3.10	33%
PFL	0.238	0.0570	4.17	
PFL (squared)	−0.00253	0.000621	4.07	56%
UNG	0.665	0.235	2.83	
UNG (squared)	−0.0386	0.0126	3.06	68%

Table 6. Sorted Spearman Rank correlation coefficients showing the relationship between 12 habitat variates and the number of young Tawny Owls fledged from 30 pairs (* = $p < 0.05$).

			Total young fledged				
All years		High vole years		Declining vole years		Low vole years	
PFL	0.45*	PFL	0.39*	PFL	0.46*	MAT	0.43*
UNG	0.29	UNG	0.37*	EST	0.45*	DIV	0.36
PTK	0.27	PTK	0.37*	UNG	0.20	THK	0.18
EST	0.28	ALT	0.30	DIV	0.08	GRZ	−0.01
DIV	0.26	DIV	0.20	WAT	0.08	EST	−0.10
ALT	0.13	EDG	0.15	PTK	0.05	PTK	−0.13
EDG	0.03	EST	0.07	ALT	0.05	WAT	−0.13
NND	−0.15	GRZ	−0.04	EDG	0.01	UNG	−0.14
GRZ	−0.22	NND	−0.05	NND	−0.03	ALT	−0.17
WAT	−0.22	MAT	−0.09	THK	−0.27	PFL	−0.38*
MAT	−0.27	THK	−0.34	GRZ	−0.31	EDG	−0.41*
THK	−0.39	WAT	−0.35	MAT	−0.46*	NND	−0.46*

to poor vole habitat (THK). However, in low vole years these habitats were less important, as the most successful territories were closer together (NND), and with a spatially more diverse habitat (DIV).

Relationship between habitat composition and Tawny Owl density

A Spearman Rank correlation matrix of untransformed habitat data was used to try and explain variations in Tawny Owl density (NND). Variates which showed significant positive correlations with NND were; UNG, EDG and WAT; while MAT and DIV showed a significant negative correlation. DIV can be considered a measure of the spatial diversity of a territory, and showed the most significant relationship with NND (Figure 6). Therefore, it was considered the most useful explanation of Tawny Owl density because all territories contained some DIV, unlike other significant variates such as MAT, UNG and WAT which had some zero values. It should be noted, however, that DIV is not totally independent of

NND. Altitude, did not have a significant effect either on spacing or productivity of Tawny Owls.

Discussion

There have been few studies which have quantified the effect of habitat composition on the density and productivity of any species of owl, although Newton et al. (1979) were the first to use a similar type of analysis for Sparrowhawks Accipiter nisus. Southern (1970) showed in lowland broadleaved woodland that the density of Tawny Owls and the number of young produced were greater in closed-canopy woodland. Here the owls fed largely on Bank Voles Clethrionomys glareolus and Wood Mice Apodemus sylvaticus which were both more abundant in woodland than in other habitats. Hirons (1985) showed a similar density effect in farmland, with the smallest Tawny Owl territories containing the most closed-canopy woodland.

The present study investigates what habitat characteristics were important for Tawny Owls in a conifer forest, where the main food throughout the year was Field Voles (Petty 1987a, Petty unpublished data). This rodent feeds largely on grasses, and was most abundant in ungrazed (by domestic animals) grassy habitats such as young plantations and grassland. It was less common in heavily grazed areas or in closed-canopy conifer forests with little ground vegetation (Hansson 1971; Corbet & Southern 1977; Charles 1981). Tawny Owls are largely 'sit and wait' hunters, and forest edges along-side recently clear-felled areas provide ample perches from which to hunt. Tawny Owls also require trees for roosting and breeding. Martin (1986) showed that Tawny Owls' vision was similar to that of humans, and so the structure of the forest must be open enough to allow the birds good access below the tree canopy at night.

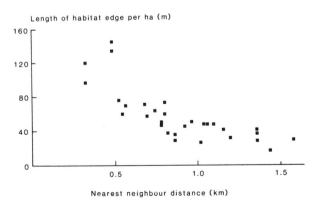

Length of habitat edge per ha (m)

Nearest neighbour distance (km)

Figure 6. Relationship between nearest neighbour distance and length of habitat edge per ha ($r_s = -0.840$, df = 28, $p < 0.001$) in 30 Tawny Owl nesting territories.

It was important to consider these opposing habitat requirements of owls, on one hand for foraging while on the other for roosting/nesting.

Therefore, it was not surprising that no single habitat measurement explained a large proportion of the variation in the productivity of Tawny Owl territories. Many habitat categories showed a non-linear relationship with productivity. For instance, the effects of pre-felling plantations (PFL) increased to an optimum and then declined (Figure 5). PFL crops provide Tawny Owls with roosting and nesting habitat but with little food, whereas UNG, EST and PTK provide the opposite. Therefore, it was reasonable to expect productivity to decline when too little or too much of either of these habitats was present, and it was not surprising that much of the variability in the productivity of territories was better explained by a curvilinear model incorporating a number of habitat variables.

The importance of different habitats for Tawny Owls changed in relation to the vole population levels. In years with high vole numbers, the best vole habitats showed the highest correlation with productivity, whereas in low vole years habitat diversity appeared to be more important, presumably because Tawny Owls were better able to find alternative food here. Songbirds, for instance, are more numerous in complex as opposed to structurally simple habitats (Newton 1986b; French *et al.* 1986), and Frogs *Rana temporaria* would be available along wetland edges. Both birds and frogs are taken by Tawny Owls in spring and summer (Petty 1987a).

Over the seven years, the spacing of owls (NND) was unrelated to the number of chicks produced (Table 6). This was true also in the three high and two declining vole years; but in the two low vole years, when a density-dependent effect might have been expected, the opposite occurred. The pairs closest together produced the most chicks. The spacing of owls (NND) was strongly associated with habitat diversity (DIV), which also suggested that it might be the presence of alternative food that enabled the owls to raise more chicks, in these smaller but structurally richer territories which were better able to maintain Tawny Owls when Field Voles were scarce.

Altitude, which is often related to raptor density, for instance Sparrowhawk (Newton 1979; Newton 1986a), appeared to have little effect on the spacing of Tawny Owls in this study. However, the lower altitude range had been truncated by the flooding of Kielder Water.

The object of this analysis was twofold; to determine how (1) the flooding of 1,000 ha for Kielder Water and (2) the felling and establishment of second generation forest, affected both the production and density of Tawny Owls. The creation of Kielder Water can be viewed as having a negative effect on productivity and also density (Table 4). It may also have resulted in the complete loss of some territories at the south-eastern end of the reservoir. In contrast, restocking appeared to be beneficial, particularly in good vole years, by providing foraging habitats which are surrounded by forest from which the owls can hunt. The density of Tawny Owls was related to the spatial diversity of the habitat. Thus management practices which provide more edges per unit of habitat, such as small opposed to large felled sites, are in the long-term likely to sustain a greater density of Tawny Owls.

While this analysis suggests that habitat composition plays an important role in the long-term productivity of owl territories, other factors not considered here may also be involved. The most obvious is the quality of the adult owls, which are relatively long-lived, and this aspect will be investigated elsewhere.

Acknowledgements

I am grateful to the Forestry Commission staff in Kielder Forest for their co-operation. I wish to thank David Anderson who helped with the field work, particularly in 1987, John Williams for drawing the figures, Linda Petty and Diane Chadwick for typing and altering numerous drafts, and Phil Ratcliffe, Ian Newton and two anonymous referees for their helpful comments on an earlier draft of this paper.

References

Anon 1984. *Census of woodlands and trees 1979–1982, Great Britain.* Forestry Commission, Edinburgh.
Charles, W.N. 1981. Abundance of the Field Vole *Microtus agrestis* in conifer plantations. In, *Forest and woodland ecology.* Ed.by F.T.Last, 135–137. Institute of Terrestrial Ecology, Cambridge.
Corbet, G.B. & Southern, H.N. 1977. *The handbook of British mammals.* Blackwell Scientific Publications, Oxford.
Cramp, S. (ed.) 1985. *The birds of the Western Palearctic.* Volume IV. Oxford University Press, Oxford.
French, D.D., Jenkins, D. & Conroy, J.W.H. 1986. Guidelines for managing woods in Aberdeenshire for songbirds. In: *Trees and wildlife in the Scottish uplands.* Ed. by D. Jenkins, 129–143. Institute of Terrestrial Ecology, Huntingdon.
Hansson, L. 1971. Habitat, food and population dynamics of the Field Vole *Microtus agrestis* (L.) in south Sweden. *Viltrevy, 8:* 267–378.
Hirons, G.J.M. 1985. The effects of territorial behaviour on the stability and dispersion of Tawny Owl (*Strix aluco*) populations. *Journal of Zoology, London (B), 1:* 21–48.
Martin, G.R. 1986. Sensory capacity and the nocturnal habits of owls *Strigiformes. Ibis, 128:* 266–277.
Newton, I. 1979. *Population ecology of raptors.* Poyser, Berkhamsted.

Newton, I. 1983. Birds and forestry. In, *Forestry and conservation.* Ed. by E. H. M. Harris, 21–30. Royal Forestry Society, Tring.

Newton, I. 1986a. *The sparrowhawk.* Poyser, Berkhamsted.

Newton, I. 1986b. Principals underlying bird numbers in Scottish woodlands. In: *Trees and wildlife in the Scottish uplands.* Ed. by D.Jenkins, 121–128. Institute of Terrestrial Ecology, Huntingdon.

Newton, I., Marquiss, M. & Moss, D. 1979. Habitat, female age, organo-chlorine compounds and breeding of European sparrowhawks. *Journal of Applied Ecology*, 16: 777–793.

Newton, I., Marquiss, M, Weir, D.N. & Moss, D. 1977. Spacing of sparrowhawk nesting territories. *Journal of Animal Ecology, 46*: 425–441.

Petty, S.J. 1987a. Breeding of Tawny Owls (*Strix aluco*) in relation to their food supply in an upland forest. In: *Breeding and management in birds of prey.* Ed. by D. J. Hill, 167–179. University of Bristol, Bristol.

Petty, S.J. 1987b. The design and use of a nestbox for Tawny Owls (*Strix aluco*) in upland forests. *Quarterly Journal of Forestry, 81*: 103–109.

Petty, S.J. 1988. The management of raptors in upland forests. *In*: *Wildlife management in forests.* Ed. by D. C. Jardine, 7–23. Institute of Chartered Foresters, Edinburgh.

Ratcliffe, P.R. & Petty, S.J. 1986. The management of commercial forests for wildlife. In: *Trees and wildlife in the Scottish uplands.* Ed. by D. Jenkins, 117–187. Institute of Terrestrial Ecology, Huntingdon.

Southern, H.N. 1970. The natural control of a population of Tawny Owls (*Strix aluco*). *Journal of Zoology, 162*: 197–285.

The ecology and conservation of the Eagle Owl *Bubo bubo* in Murcia, south-east Spain

J.E. Martinez, M.A. Sanchez, D. Carmona, J.A. Sanchez, A. Ortuño and R. Martinez

Martinez, J.E., Sanchez, M.A., Carmona, D., Sanchez, J.A., Ortuño, A. & Martinez, R. 1992. The ecology and conservation of the Eagle Owl *Bubo bubo* in Murcia, south-east Spain. *In*: *The ecology and conservation of European owls*, ed. by C.A. Galbraith, I.R. Taylor and S. Percival, 84-88. Peterborough, Joint Nature Conservation Committee. (UK Nature Conservation, No. 5.)

Aspects of the ecology of the Eagle Owl *Bubo bubo* were studied on Murcia, south-east Spain from 1987 to 1989. A total of 142 pairs were located and the total population estimated to be 200–250 pairs. Pairs were distributed throughout but especially on areas of scrub and non-irrigated agriculture.

The number of young fledged per pair was 2.72, 3.21 and 2.28 in 1987, 1988 and 1989 respectively. Following an outbreak of a new viral disease (NHV) in the Rabbit *Oryctolagus cuniculus* population in 1988, the number of pairs of owls that nested in 1989 fell by 48.4% and population productivity fell by 60.2%.

Mammals were the most important items in the diet constituting 93.6% of biomass. Rabbits made up 80.6% of total dietary biomass.

The main causes of recorded death were shooting, electrocution and nest robbing.

J.E. Martinez, M.A. Sanchez, D. Carmona, J.A. Sanchez, A. Ortuño and R.Martinez, Asociacion Fauna Silvestre, c/Jose Maluquer 8, 2-B, Murcia, Spain

Introduction

Previous studies of the Iberian subspecies of the Eagle Owl *Bubo bubo* have dealt with food requirements (Hiraldo *et al.* 1975; 1976; Perez Mellado 1980), non-natural mortality (Gonzalez *et al.* 1979; Hernandez 1989) and other aspects of ecology (Donazar 1986).

No studies have specifically been carried out in south-east Spain therefore the only information available is from general studies (Sanchez & Carmona 1986).

In this paper we provide information on distribution, habitat selection, breeding and diet of the Eagle Owl in south-east Spain.

Study area

The study area was the Comunidad Autonoma of Murcia, an area of 11,317 km² (Figure 1). The region is mountainous with parallel ridges running W-SW to E-NE rising to 2,000 m a.s.l. Cliffs are numerous and the climate is typically Mediterranean. Rainfall is low, around 270 mm per annum and concentrated within a few days, mostly in autumn.

About 25% of the land surface is covered in scrub vegetation and 15% in woodland mostly of *Pinus*

halepensis. Agricultural land comprises about 50% of the land surface; 43% non irrigated and 13% irrigated.

The human population is approximately 1 million with a mean density of approximately 88 inhabitants per km².

Status, distribution and density

A total of 142 pairs of Eagle Owls were located and the total population (based on area) of the region was estimated to be 200–250 pairs. The population is therefore of significant size compared with other European populations (Mikkola 1983; Cramp & Simmons 1985).

Eagle Owls were found throughout the whole region, following the regular distribution of cliffs (Figure 2).

Mean density over the region was estimated to be 1 per 50 km². The mean distance between nests was estimated for two subpopulations, using the nearest neighbour method (Clark & Evans 1954 in Urios 1986). For one of these populations mean nearest neighbour distance was 5.1 km (range 2.6–6.5 km, n=8) and for the other, 4.2 km (range 2.3–10.6 km, n=11)

Figure 1. Study area.

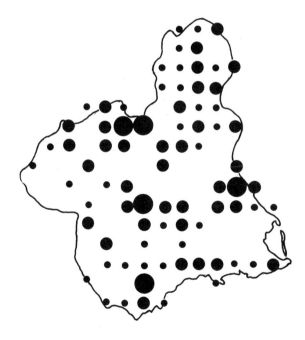

Symbols used:

●   1 pair 100 km^2

⬤   2–3 pairs 100 km^2

⬤   4 or more pairs 100 km^2

Figure 2. *Bubo bubo* distribution and density in Murcia (south-east Spain).

Habitat selection

Eagle Owl nesting density was examined in relation to three aspects of habitat: relief, vegetation and human influence, in an area of 1,300 km^2 of the coastal mountains in which 20 pairs of owls nested. The area was divided into 52 cells each of 25 km^2. Owl density and measures of the habitat variables were obtained for each cell.

An index of relief was calculated for each cell according to the following relationship.

Index = (Cx + Cy/40) + (h max – h min)/25

Cx = No of contour lines (40m altitudinal separation) cut by a line drawn vertically across the mid point of the cell.
Cy = As above but a horizontal line.
h max = maximum altitude within cell.
h min = minimum altitude within cell.
25 = surface area of each cell in km^2.

There was a highly significant positive relationship between owl density and the index of relief ($p < 0.01$).

The analysis of vegetation was based on agricultural management maps of 1 : 50,000 scale (Agriculture Ministry 1984). A circular plot of 25 km radius was drawn round the nest of each pair (M = 20) and the area of each of the following vegetation types was calculated: 1) forest, 2) non irrigated agricultural land, 3) scrub, and 4) other vegetation types. Values were then compared with the area of each of these habitat types within the area as a whole. Significant associations were found between Eagle Owl distribution and the presence of scrub and non irrigated agricultural land but not with the other two habitat types. A similar association with those habitats has been demonstrated by Blondel & Badan (1976).

Human influence on distribution was examined by comparing the Eagle Owl distribution with human population density and the density of roads. No relationship was found with road density but there was a significant negative relationship with human population density.

Breeding

Average laying date was 22 January in 1987 (n = 12) and 18 January in 1988 (n = 17) with a total range of only 14–28 January. This was similar to laying dates in south-east France (Blondel & Badan 1976) but almost a month earlier than in north and north-east Spain (Donazar 1984; 1986; Real *et al.* 1985) and earlier than central and northern Europe (Mikkola 1983; Olsson 1979).

In 1987, 63.1% of pairs laid eggs (n = 19), in 1988, 89.4% laid (n = 19) and in 1989, 41% laid (n = 17). Mean number of young fledged per pair that laid was 2.7 in 1987 (n = 11), 3.2 in 1988 (n = 14) and 2.8 in 1989 (n = 17).

Fledging success was similar or slightly higher than that recorded in France; 2.69/pair (Blondel & Badan 1976), 2.2/pair (Mikkola 1983), 2.04/pair (Cugnasse 1983).

Productivity, expressed as the number of young averaged over all pairs in the population, including those that did not lay, was 1.6 in 1987 (n = 19) 2.4 in 1988 (n = 19) and 0.9 in 1989 (n = 17). Differences between 1987 and 1988, and 1988 and 1989 were significant (matched pairs test, in both cases $p < 0.01$).

In October 1988 a new viral disease (NHV) affected the Rabbit *Oryctolagus cuniculus* population of south-east Spain producing high mortality which approached 100% in some areas. The effects of this drastic reduction in the abundance of the owl's main prey species on breeding performance 1989 were severe. Compared with 1988, the number of pairs that laid eggs fell by 48.4% and productivity (see above) fell by 60.2% (see data above). This demonstrates the vulnerability of a species so dependent on a single prey species.

Feeding ecology

Diet was determined for 11 pairs in 1986 and 1987 by the analysis of pellets taken from the nest. This method tends to underestimate the importance of large prey items (Olsson 1979). A total of 1398 items were identified and subdivided into three periods: 1) 15 October–January, pre-breeding and laying, 2) February–May, incubation and chicks and 3) May–October 14, fledging and post fledging dependence (Figure 8). Biomass of prey was calculated from weights given by Van de Brick & Barrvel (1971) and Burton & Arnold (1975).

Mammals constituted 93.6% of diet. Rabbits made up 53.6% of items and 80.6% of biomass. Birds made up 18.7% of items and 6.1% of biomass. Of these, Red-legged Partridge *Alectoris rufa* was the most important species contributing 2% of the

biomass. Insects and reptiles each contributed 1% of total biomass. (Figure 3, Table 1). Rabbits were more important in the diet during period 2, when the chicks were being fed in the nests. The most

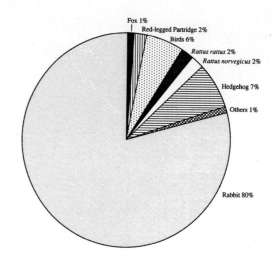

Figure 3. Diet of *Bubo bubo*. Percentage of biomass by species.

Table 1. *Bubo bubo* prey items in Murcia (1986–89).

	No.	% Numbers	% Biomass
INSECTS			
Coleoptera	55	3.93%	0.00%
REPTILES			
Testudo graeca	1	0.07%	0.08%
Lacerta lepida	2	0.14%	0.02%
MAMMALS			
Oryctolagus cuniculus	750	53.60%	80.58%
Mus musculus	21	1.50%	0.04%
Rattus rattus	128	9.14%	2.47%
Rattus norvegicus	76	5.43%	2.28%
Elyomis quercinus	3	0.21%	0.02%
Erinaceus algirus	79	5.64%	7.00%
Arricola sapidus	7	0.50%	0.11%
Mustela nivalis	6	0.42%	0.01%
Myotis myotis	6	0.42%	0.01%
Vulpes vulpes	2	0.14%	1.07%
Total mammals	1078	76.50%	93.58%
BIRDS			
Columba sp.	41	2.93%	0.88%
Turdus merula	33	2.36%	0.35%
Corvus corax	2	0.14%	0.17%
Pyrrhocorax pyrrhocorax	18	1.28%	0.38%
Corvus monedula	7	0.50%	0.18%
Alectoris rufa	41	2.93%	1.98%
Picus viridis	18	1.28%	0.38%
Larus argentatus	3	0.21%	0.19%
Sturnus unicolor	13	0.92%	0.06%
Oenanthe leucura	1	0.07%	0.00%
Upupa epops	4	0.28%	0.08%
Falco tinnunculus	17	1.21%	0.36%
Buteo buteo	1	0.07%	0.07%
Accipiter nisus	1	0.07%	0.07%
Athene noctua	21	1.50%	0.22%
Birds indetermined	41	2.92%	0.74%
Total birds	262	18.67%	6.06%
TOTALS	1,398	100.00%	100.00%

frequently taken prey items, constituting 79.8% of an item, were between 0.5 and 1.2 kg (Figure 4).

The very high dependence on mammals in Murcia exceeds that recorded previously in all other parts of Europe (80% Cugnasse 1983; 79% and 64% Perez Mellado 1980; 69% Mikkola 1983; 73% Blondel & Badan 1976; 85% Real *et al.* 1985; 62% Thiollay 1968). The importance of the Rabbit as the main prey species in Mediterranean areas (Figure 5) has been reported in other studies (Donazar 1986;

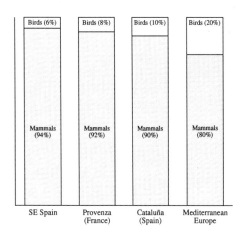

Figure 6. Diet composition of *Bubo bubo* in four locations of Mediterranean Europe.

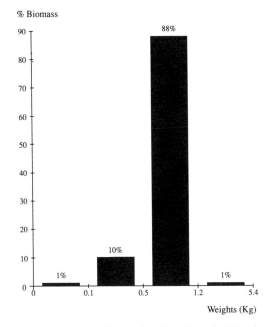

Figure 4. Prey size selection by *Bubo bubo* in Murcia.

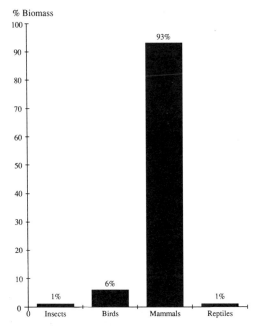

Figure 5. Diet of *Bubo bubo*. Classification by taxonomic groups.

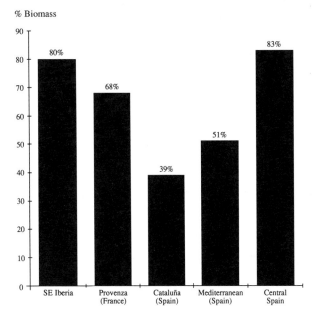

Figure 7. Percentage of *Oryctolagus cuniculus* in the diet of *Bubo bubo* in different areas of Europe.

Hiraldo *et al.* 1975, Blondel & Badan 1976), where microtine voles are absent.

Between 1970 and 1989 the causes of death were known for 84 Eagle Owls in the study area. Shooting was the main cause (55.9%) followed by nest robbing and electrocution (17.8% each). Poisoning and trapping (2.3% each) were not important but may have been underestimated (Figure 9).

Shooting has been shown to be important elsewhere in Spain (Hernandez 1989) but of little significance elsewhere, (Olsson 1979; Wickl 1979). Electrocution is an important factor in other European countries (Choussy 1971; Wickl 1979).

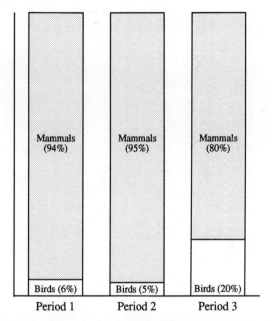

Figure 8. Seasonal evolution of the diet of *Bubo bubo* in Murcia.

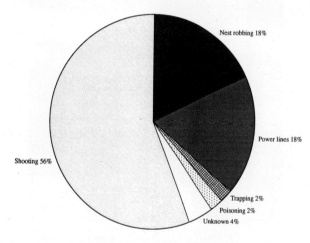

Figure 9. Non natural mortality of *Bubo bubo* in Murcia.

Acknowledgements

We thank the following: A. Martinez, A. Pagan, M. Padin, I. Padin, J. Linares, P. Sanchez, A. Zamora, P. Cortes, A. Torres, Toni, Encarnita, Carana and Forest Guards, D. Reales and J. Garcia.

References

Blondel, J. & Badan, O. 1976. La biologie du Hibou Grand Duc en Provence. *Nos Oiseaux, 33*: 189–219.

Burton, J.A. & Arnold, E.N. 1982. *Guia de campo de los reptiles y anfibios de Espana y Europa.* Omega. Barcelona.

Cramp, S. & Simmons, R. 1985. *The birds of the Western Palearctic. Vol. IV.* Oxford University Press, Oxford.

Cugnasse, J.M. 1983. Contribution a l'etude de Hibou Grand Duc *(Bubo bubo)* dans le sud Massif Central. *Nos Oiseaux, 37*: 117–128.

Choussy, F. 1971. Etude d'une population de Grand Duc *(Bubo bubo)* dans le Massif Central. *Nos Oiseaux, 31*: 37–56.

Donazar, J.A Ceballos, O. 1984. Alqunos datos sobre status, distribucion y alimentacion del Buho real *(Bubo bubo)* en Navarra. *Rapinayres*, 2: 246–254.

Donazar, J.A. 1986. *El Buho real (Bubo bubo, L) en Navarra: poblacion, analisis de la distribucion, ecologia trofica reproduccion y crecimiento.* Tesis doctoral. Inedito. Universidad de Navarra. Pamplona.

Gonzalez, J.L., Lobón, J., Gonzalez, L.M. & Palacios, F. 1979. Datos sobre la evolucion de la mortalidad no natural del Buho real *(Bubo bubo)* en Espana durante el periodo 1972–80. *Boletin de la Estacion Central de Ecologia.* ICONA: 63–65.

Hernandez, M. 1989. Mortalidad del Buho real en Espana. *Quercus, 40*: 24–25.

Hiraldo, F., Andrada, J. & Parreño, F. 1975. Diet of the Eagle Owl *(Bubo bubo)* in Mediterranean Spain. *Donana Acta Vertebrata, 2*: 161–177.

Hiraldo, F., Parreño, F., Andrada, V. & Amores, F. 1976. Variations in the food habits of the European Eagle Owl *(Bubo bubo)*. *Donana Acta Vertebrata, 3*: 137–156

Mikkola, H. 1983. *Owls of Europe.* Poyser, Calton.

Ministerio de Agricultura, Pesca Y Alimentacion, 1984. Mapa de cultivos y aprovechamientos de la provincia de Murcia, Madrid.

Olsson, V. 1979. Studies on a population of Eagle Owl in southeast Sweden. *Viltrevy, 11*: 1–99.

Perez Mellado, V. 1980. Alimentacion del Buho real *(Bubo bubo)* en Espana Central. *Ardeola, 25*: 93–112.

Real, J., Galobart, A. & Fernandez, J. 1985. Estudi preliminar d'una poblacio de Duc *(Bubo bubo)* al Valles i Bages. *Medi natural del Valles*, 175–187.

Sanchez, M.A. & Carmona, D. 1986. *Status y distribucion de las avesde presa (O falconiformes y O Estrigiformes) en Murcia SE de Espana.* V Conferencia International sobre rapaces Mediterraneas. Evora. Portugal. Inedito.

Thiollay, J.M. 1968. Essai sur les rapaces du Midi de la France. Distribution-ecologie, hibou Grand-Duc *(Bubo bubo)*. *Alauda, 37*: 15–27.

Urios, V. 1986. Biologia, requerimientos ecologicos y relaciones interespecificas del Aquila real *(Aquila chrysaetos homeyeri)* y del Aquila perdicera *(Hieraetus fasciatus fasciatus)* en la provincia de Valencia. Tesina de Licenciatura. Inédito. Universidad de Valencia.

Van de Brick, F. & Barruel, P. 1971. *Guia de campo de los mamiferos salvajes de Europa occidental.* Ediciones Omega. Barcelona.

Wickl, K. 1979. Der Uhu *(Bubo bubo)* in Bayern. *Garmicher Vogelkundliche Berichte, 6*: 1–47.

A guide to age determination of Tawny Owl *Strix aluco*

by S.J. Petty

Petty, S.J. 1992. A guide to age determination of Tawny Owl *Strix aluco*. In: *The ecology and conservation of European owls*, ed. by C.A. Galbraith, I.R.Taylor and S. Percival, 89-91. Peterborough, Joint Nature Conservation Committee. (UK Nature Conservation, No. 5.)

S.J. Petty, Forestry Commission, Wildlife & Conservation Research Branch, Ardentinny, Dunoon, Argyll PA23 8TS.

This method, for the age determination of Tawny Owls *Strix aluco* has been developed from known-aged individuals (ringed as chicks) and colour dyed adults (Figure 1) in a study of Tawny Owls in conifer forests in Northumberland and Argyll.

Tawny Owls which breed do not start to moult their wing and tail features until after the chicks have fledged. Non-breeding Tawny Owls commence their moult earlier. Moult of primary and secondary feathers usually occurs from May to October.

In each wing Tawny Owls have ten primary feathers which here are numbered from P1 (inner) to P10 (outer wing) and 13 secondary feathers numbered from S1 (outer) to S13 (inner wing).

Tawny Owls have distinct juvenile and adult primary/secondary feathers. Juvenile feathers have a broken or thin terminal band (Figure 4, all feathers except P5 and P6), and may have a different colour and/or pattern to juvenile feathers.

Tawny Owls have a complex moult with some juvenile feathers retained until the third annual moult (in the fourth year of life) after which none remain (Figure 2).

Tawny Owls (male and female) which are caught while breeding and before moulting can be divided into four age categories. Examples of these are shown in Figures 3–6. It is unusual for moult sequences to be the same in both wings, so both need to be checked.

This method of age determination cannot be used for birds kept in captivity. Care should also be taken in applying it to other wild Tawny Owl populations, as differences in food supply and reproduction may affect moulting patterns.

To validate the method, the age of adult Tawny Owls caught while breeding was estimated in the field using the above criteria (Figures 3–6). Estimated ages were then compared at a later date to the actual age of

102 Tawny Owls which had been ringed as chicks (Table 1). Birds over five years of age were excluded as all these were classified as three-plus-year old. All birds estimated as one- or two-years olds were correctly aged. Ten out of 51 (19.6%) three-year olds were misclassified as three-plus-year olds because all their juvenile feathers had been replaced at the second moult. More males (31.2%) were misclassified compared to females (14.3%) but this difference was not significant (chi-squared = 1.07, df = 1, p > 0.05). All four-year olds were classified as three-plus-year olds, none having juvenile feathers retained beyond their third moult.

Table 1. Actual age of male (M) and female (F) Tawny Owls compared to their estimated age using the criteria illustrated by Figures 3–6.

Actual age	sex	Estimated age			
		1	*2*	*3*	*3+*
1	M	6	0	0	0
	F	5	0	0	0
	M + F	11	0	0	0
2	M	0	7	0	0
	F	0	11	0	0
	M + F	0	18	0	0
3	M	0	0	11	5
	F	0	0	30	5
	M + F	0	0	41	10
4	M	0	0	0	9
	F	0	0	0	13
	M + F	0	0	0	22

Acknowledgements

I am very grateful to David Anderson who helped me catch and dye many Tawny Owls, Phil Ratcliffe and Kevin Baker for their helpful comments on the text, John Williams for drawing the figure and Diane Chadwick for typing numerous drafts.

Figure 1. Moult patterns in Tawny Owls have been investigated by dyeing (with picric acid) primary and secondary feathers of breeding adults prior to moult commencing, and then recapturing them a year later. This is typical of a bird that has successfully reared chicks, replacing only S1, S10 and S11 during the year. Non-breeding birds and those which have failed in a breeding attempt moult considerably more feathers.

Figure 3. Yearling, prior to its first moult in its second year of life. All the primary and secondary feathers are juvenile. This bird also has some juvenile (white-tipped) greater coverts, these are completely replaced by adult feathers before November/December in the first year of life. Some birds may have moulted all their juvenile greater coverts as early as September/October.

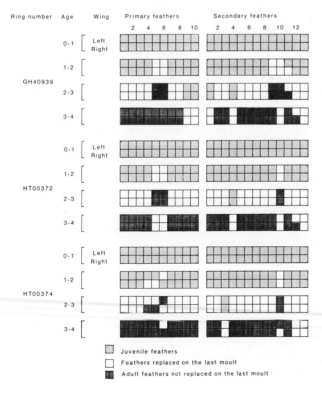

Figure 2. Wing feather moult of three female Tawny Owls through their first three annual moult cycles. Age 0–1 is up to the start of the first moult (in the second year of life), age 1–2 is from after the first moult until before the second moult (in the third year of life), etc.

90

Figure 4. Two-year old bird, prior to its second moult in its third year of life. During the first moult only two primary feathers (P5 and P6) were replaced by adult feathers, the rest are juvenile feathers. Primary moult usually commences with P5 and P6 and then progresses both outwards to P10 and inwards to P1. None of the outer secondary feathers were moulted, S1-S10 are juvenile. The number of wing feathers replaced at the first and subsequent moult varies and depends on food supply, which in turn affects reproductive performance. Fewer feathers are moulted in years when chicks are reared successfully (Petty unpublished data).

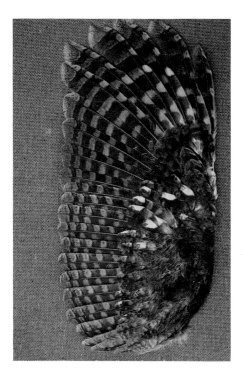

Figure 6. Three-plus-year old bird. This bird has no juvenile primary or secondary feathers. Its minimum age is four years, prior to its fourth moult in its fifth year of life. Adult feathers can be retained through three or more annual moult cycles, so beware of misclassifying very worn feathers as juvenile.

Figure 5. Three-year old bird, prior to its third moult in its fourth year of life. All the primary feathers have now been replaced by adult feathers. Juvenile secondary feathers are still retained at S1, S4, S7, S8 and S11. Virtually all three-year old birds retain at least one juvenile feather, S4 or occasionally S1 and/or S7 are usually the last to be replaced (Figure 2).

To establish the age of two-year and three-year old birds it is necessary to determine whether one (two-year old) or two (three-year old) generations of adult feathers are present. To do this, turn the wing over and look at the white area at the base of the primary and secondary feathers. It is best to blow or move the underwing coverts to one side. Feathers that were replaced on the last moult have a pink/brown (brown feathers) wash to this white area, while feathers replaced on the last but one moult do not (white feathers). Therefore, birds with a mixture of white and brown feathers are three-year olds, while two-year olds have brown feathers only. In poor light and sunny conditions it may not be possible to detect these two types of adult feathers. Overcast but reasonably bright conditions are the best. Retained juvenile feathers are usually very well worn in three-year old birds.

Radio-tracking Tawny Owls *Strix aluco* in an upland spruce forest

By J.P. Gill

Gill, J.P. 1992 Radio-tracking Tawny Owls Strix aluco in an upland spruce forest. *In: The ecology and conservation of European owls*, ed. by C.A. Galbraith, I.R. Taylor and S. Percival, 92-93 Peterborough, Joint Nature Conservation Committee (UK Nature Conservation, No. 5.)

J.P. Gill, Forestry Commission, Wildlife and Conservation Research Branch, Ardentinny, Dunoon, Argyll PA23 8TS

This poster summarises a study reported in Gill (1989), further details of which will appear in Gill, Thirgood and Petty (in prep).

1. Radio telemetry was used to track the movements of five male, seven female and seven juvenile Tawny Owls in an upland spruce forest in Argyll from April to August 1989 (Gill 1989). Both back- and tail-mounted radio-tags were fitted to birds caught at nestboxes. Analysis of results was by a method which uses a kernel to smooth the distribution of an animal's locations (Worton 1989).

2. The use of Territory 200 by the resident female was compared with her use of the same territory in 1986 (Figure 1a and 1b). In 1989 the young conifers to the east were used more than in 1986, and a smaller proportion of locations were in broadleaved habitat. Her core area was less tightly clustered around the nestbox and she was located over a larger area in 1989 than in 1986 (Figure 1b).

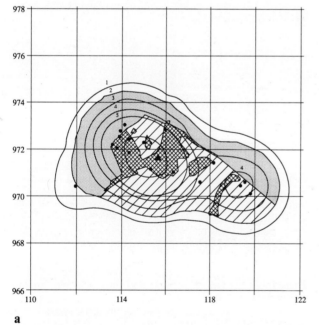

a

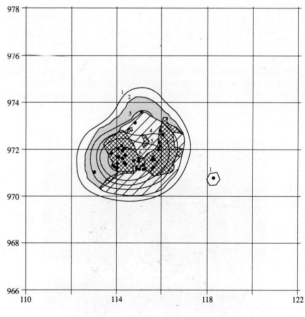

High conifer forest

Other conifer forest

Broadleaved areas

Open habitat

▲ Nest box location

• Location of birds recorded by radio tracking

Figure 1. Habitat use by the same female in Territory 200 in 1989 (Figure 1a) and 1986 (Figure 1b). The contour labelled 1 corresponds to a probability of 0.95 of locating the subject. Similarly 2, 3, 4, 5 and 6 correspond to 0.9, 0.8, 0.6, 0.5 and 0.25 respectively. Core areas could be identified within the 0.6 and 0.8 contours. Each radio-tracking location of the owl is shown by a dot. Grid lines are separated by 200 m and lie within the 10 km square NS.

3. In both years food availability was low, and juveniles died within a few days after fledging. Both studies revealed a preference by the female for broadleaved areas, but insufficient data were collected from the male for his habitat preferences to be analysed. In 1986 the male was not tagged, and in 1989 the feather to which a tag was attached was moulted within one month.

4. The female in Territory 155 was located most often in conifer crops planted before 1953 (high conifer forest). She also used younger conifers, and was found hunting along the forest/moorland boundary and over open areas where conifer crops which had been clear-felled and replanted in 1986, at the northern end of her territory (Figure 2). The locations in the north-west of Figure 2 were recorded during a single night. This excursion greatly increased her home range. Extensive clearfelling was occurring at this time in her core area.

5. The "kernel" model was helpful in preliminary analysis of radio-tracking data. Back-mounted radios were more suitable for Tawny Owls as some of the feathers supporting the tail-mounted radios were shed prematurely.

6. Most of the radio-tagged owls showed significant selection for broadleaved areas and older conifer crops. In order to create good quality Tawny Owl habitat in second-generation spruce forests both of these habitats should be well represented.

Acknowledgements

I gratefully acknowledge the permission of the Forestry Commission to conduct the study on their land, and the efforts of Steve Petty in setting up the project. I thank George Gates for preparing the figures and captions, Steve Petty and Phil Ratcliffe for making helpful suggestions and Diane Chadwick for typing the manuscript.

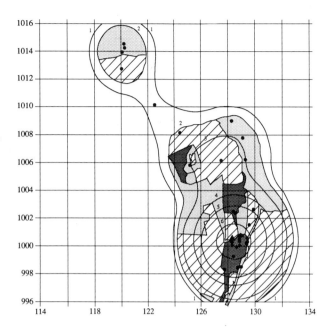

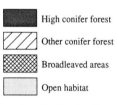 High conifer forest

Other conifer forest

Broadleaved areas

Open habitat

▲ Nest box location

• Location of birds recorded by radio tracking

Figure 2. Habitat use by the female in Territory 155. Contours, symbols and habitat types are as in Figure 1. The line marked 1000 separates 10 km squares NS to the south from NN to the north.

References

Gill, J.P. 1989. *Some details of the breeding biology and ecology of tawny owls (Strix aluco), as revealed by supplemental feeding and radio telemetry in Cowal Forest in Argyll, Scotland.* MSc Thesis, Imperial College, University of London.

Worton, B.J. 1989. Kernel methods for estimating the utilisation distribution in home range studies. *Ecology*, 70: 164–168.

Road deaths of Westphalian owls: methodological problems, influence of road type and possible effects on population levels

H. Illner

Illner, H. 1992. Road deaths of Westphalian owls: methodological problems, influence of road type and possible effects on population levels. *In: The ecology and conservation of European owls*, ed. by C.A. Galbraith, I.R. Taylor and S. Percival, 94-100. Peterborough, Joint Nature Conservation Committee. (UK Nature Conservation, No. 5.)

1. From 1974 to 1986, 74 Barn Owl *Tyto alba*, 46 Little Owl *Athene noctua*, 20 Long-eared Owl *Asio otus* and 11 Tawny Owl *Strix aluco* were found dead or injured in an area of 125 km^2. The majority of finds resulted from talks with farmers, during extensive studies of breeding biology and annual mapping of owl territories. If the dead or injured owl could not be examined by the author, detailed information was sought from the finder. The lowest proportion of finds checked personally was for the Barn Owl (46%). Proximate causes of death were determined as far as possible.

2. Different information sources gave different ratios in the cause of death both within and between species. Some differences were easily explained through special searching activities of the author or the life-style of the owl.

3. A comparison of different enquiry methods showed that deaths of Barn Owl and Little Owl on roads were significantly over estimated by the ringing and recovery method.

4. Between 54 and 68% of deaths of the four owl species were probably caused by man, with road deaths ranging from 25% in the Long-eared Owl to 32.5% in the Little Owl.

5. On roads with car speeds of more than 80 km/h about 21 times as many owls were killed by cars as on roads with slower traffic. The road death rate of the Little Owl was significantly greater compared with the Barn Owl if roads with fast-moving traffic passed through villages. This could be explained by this species prevalence for breeding in villages and its small home range.

6. Minimum percentages of road deaths were obtained by calculating known road deaths in relation to the known breeding populations (from intensive mapping studies). These values were, for Barn Owl 6.5%, Little Owl 1.5%, Long-eared Owl 1%, and Tawny Owl 0.5%. Subtracting first-year birds and multiplying these percentages by an estimated 'unknown' factor of owls not recorded but killed by cars, the highest road death figures were calculated for Barn Owl (6.5%) and Little Owl (4%). The high mortality of Barn and Little Owls on roads may have contributed to the long-term decrease of these two species.

H. Illner, I. Zoologisches Institut der Universität, Berliner Straße 28, D-3400 Göttingen, Federal Republic of Germany

Present address: Postfach 1537, Schottenteich 12, D-4770 Soest, Federal Republic of Germany

Introduction

Deaths of birds and other animals on roads were first investigated in Europe in the 1930s (Barnes 1936, cited in Finnis 1960; Löhrl 1950). Following the huge increase in car traffic after the Second World War, many such studies have been undertaken in Western Europe (for example Finnis 1960; Hodson 1960; Hansen 1969; Bergmann 1974; but few have contained much information about owls (Hodson & Snow 1965; Weir 1971; Uhlenhaut 1976; Schönfeld et al. 1977; Bunn et al. 1982; Bourquin 1983; Shawyer 1987).

Although the ratio of road casualties as a proportion of all causes of death has been calculated many times for owls with ringing and recovery data (for example Glue 1971; Bezzel 1982), the proportions attributed to traffic seem to be overestimated because of methodological bias. Hodson & Snow (1965), Cavé (1977), Newton (1979) and Anderson et al. (1985) have shown that the probability of finding a dead ringed bird is likely to be a function of the cause of death, geographic location of the bird and the time of year. Furthermore it is likely that the probability of reporting a found ringed bird is not constant. Some bird species such as the Barn Owl are highly prized as stuffed specimens and might therefore be reported more frequently than other species (see Weir 1971 for further discussion of higher reporting rate for uncommon species and unusual death causes).

Some attempts have been made to estimate the mortality rate on roads of songbirds through ringing juvenile birds near roads and searching for them as road victims (Dunthorn & Errington 1964) or

94

through extrapolations from breeding densities, productivity and road deaths (Hodson & Snow 1965; Smettan 1988). Only three owl studies show either road death mortality rates computed for real populations (Shawyer 1987) or published data of breeding density and road deaths in defined areas and years (Weir 1971; Schönfeld et al. 1977).

Preliminary results of an owl population study in central Westphalia (Illner 1981) summarized the cause of owl deaths from 1974 to 1979. Information on road deaths of the Barn Owl *Tyto alba*, Little Owl *Athene noctua*, Long-eared Owl *Asio otus* and Tawny Owl *Strix aluco* from 1974 to 1986 are presented. Comparisons were made of different enquiry methods and time periods, and the influence of road type and car speed on the recorded mortality rate. In conclusion the author tries to answer the question of whether road deaths may have been responsible for the long-term population decline in the four owl species (Illner 1988).

Material and methods

The material comprised 151 owls found dead or seriously injured from 1974 to 1986. Owls younger than three months were excluded as the main aim of the work was to investigate the impact of road deaths on the overall breeding population rather than its effect on birds during the immediate post-fledging period, when natural mortality is high. The finds came from different sources (Table 1). The majority of finds resulted from conversations with farmers during the authors' population studies (annual mapping of owl territories and extensive studies of owl breeding biology; (for details see Illner 1981, 1988). A few records of recoveries were initiated by telephoning or writing to the author or writing to the ringing centre after finding a ringed owl.

The study area covered 125 km², with an altitudinal range of 70–230 m. The area was mainly cereal farming (about 60%), with some woodland (about

Table 1. Sources of information in the enquiry into causes of death of owls in the study area 1974–1986; sick birds not fit for survival in the wild included. The cause of death classes are: road death (a), other deaths caused by man (b), natural causes (c), and unknown (d).

Total	Total finds in cause of death class	Recovery from the ringing centre*	Personal find	Checked information from other persons**	Unchecked information from other persons
	n	%	%	%	%
Barn Owl					
(a)	22	32	–	23	45
(b)	28	–	4	21	75
(c)	9	–	–	(44)	(56)
(d)	15	–	13	27	60
Total	74	9.5	4	25.5	61
Little Owl					
(a)	15	33	–	77	–
(b)	12	17	17	25	41
(c)	11	9	64	18	9
(d)	8	(25)	(25)	–	(50)
Total	46	21.5	24	33	21.5
Long-eared Owl					
(a)	5	–	–	(40)	(60)
(b)	6	(17)	(33)	(33)	(17)
(c)	8	–	(75)	(12.5)	(12.5)
(d)	1	–	–	–	(100)
Total	20	5	40	25	30
Tawny Owl					
(a)	3	–	(33.5)	–	(66.5)
(b)	4	–	–	(50)	(50)
(c)	2	–	–	(50)	(50)
(d)	2	–	–	(100)	–
Total	11	–	9	45.5	45.5
Grand total	151	12	15	29	44

*Information only obtainable from this source.
**Checked by personal examination of find or ring.
Percentages are given in parentheses where the number of corpses in the sample was less than ten.

5%) and grassland (about 10%). More details are given by Illner (1981, 1988).

Verbal reports of dead owls were checked by personal inspection as far as possible (overall this was achieved for 29% of the records, Table 1). If the find was no longer available the finder was questioned about the precise size, appearance and condition of the bird, the finding data and location and the finding circumstances. The same procedure was followed with persons who reported (to the ringing centre) finding owls ringed by the author. This eliminated many mistakes and inaccuracies of ringing and recovery data, especially in relation to the location and condition of the found bird. All finds for which the author could not be sure of the identification, cause of death or the location were separated out into special categories. The questionnaire avoided suggestive questions as far as possible. Dead birds found by the author or obtained from other persons (about one third of the total sample) were examined as follows: weight, other body measurements, condition (such as size of pectoral muscle), injuries and signs of diseases, stomach content, and sometimes chemical analyses of organochlorine residues through the Tierhygienisches Institut Freiburg and Tierärztliche Hochschule Hannover.

The determined causes of death have to be considered as proximate ones after Newton (1979). Because a road death can be determined as the proximate cause with little error (most were found on the road or the roadside verge and showed signs of collision; no roadside wires were present in the study area so this could not have complicated the results) by the questionnaire method, it is justifiable to list it separately in the tables of results. The other causes of

death are divided into two classes with several causes of death and a class 'Unknown'. The error of classifying into these three death-cause groups was estimated as small.

Table 2 shows that different ratios of the causes of death were found both within and between species. Some findings are easily explained. The high figures in the class "personal find" for Little Owl and Long-eared Owl correspond with high personal activity through extensive studies of breeding and winter ecology of Little Owls, and many surveys of coniferous woods, searching for Long-eared Owls and Sparrowhawks *Accipiter nisus*. This resulted in many finds of conspicuous raptor kills of Long-eared Owls. Likewise the high figures for unchecked information of dead Barn Owls are explained through the use by this species of buildings for breeding and roosting in the study area. Many were found and reported in buildings later during the breeding census

Because the number of finds (Table 1) are far below the real number of deaths and, as mentioned above, species-specific and death cause-specific finding and reporting rates are obvious, it must be concluded that the ratios of different causes of death classes (Table 3) were not representative of the study populations. An estimate of total mortality in the same manner as for the road death mortality alone (see Table 5) yielded annual adult mortality figures of about 11% for Barn Owl, 3.5% for Little Owl, 2.5% for Long-eared Owl and 1% for Tawny Owl. These figures are much lower than those of Glutz & Bauer (1980), but the relationship between the figures for each species is similar.

Road types (Table 4) were ascertained with topographical maps (1 : 25 000) and with personal

Table 2. The proportion of deaths of Barn and Little Owls attributable to road accidents recorded by different enquiry methods, 1974–1987 (1987 one find).

Enquiry method	Barn Owl		Little Owl	
	Total	Road (%)	Total	Road (%)
(1) Own research (without ringing and recovery)*	66	15 (23%)	38	9 (27%)
(2) Ringing and recovery – recoveries inside study area	8	7 (88%)	8	6 (75%)
(3) Ringing and recovery – recoveries up to 200 km.	39	17 (44%)	16	8 (50%)

The difference between the method (1) and (2) was significant both the Barn and Little Owl ($p < 0.05$, Fisher's exact test). The difference between the method (1) and (3) was significant for the Barn Owl ($p < 0.05$, χ^2 test), and tending towards significance for the Little Owl (for which samples were smaller ($p < 0.1$, χ^2 test)).
*Includes: banded owls which were found dead but were not reported by the finder to the ringing centre but to the author (Barn Owl, $n = 13$; Little Owl, $n = 15$).

knowledge of all roads in the study area. Road sections were classified into two groups; those on which cars regularly travelled at speeds greater than 80 km/h and those where car speeds were below 80 km/h on at least 99% of occasions. This was assessed by personal experience as a car driver, following other cars with speed estimation, testing maximum speed without losing control of the car and many speed estimates during fieldwork. Apart from some short new road sections, especially in rebuilt areas (nearly all with car speeds less than 80 km/h), the road system was about the same in 1974 and 1986. Car traffic volume increased each year from 1974 to 1986. Top speeds particularly increased in the mid-1980s, but even in the 1970s some car drivers exceeded 100 km/h at night even on some country roads and most drivers exceeded 100 km/h on trunk roads.

Results and discussion

A comparison of different enquiry methods (Table 2) for the two owl species with most finds showed that road deaths were significantly overestimated through the ringing and recovery method relative to the author's own research (method (1) versus (2)).

This confirms the presumptions of Hodson & Snow (1965), Glue (1971), Weir (1971) and Newton (1979) that deaths caused by man, especially road deaths, are overestimated in bird mortality studies which are primarily based on reports. It is likely that road deaths and other deaths caused by man are still overestimated in this relatively intensive study (see above) and therefore that there is an even greater bias in the ringing and recovery method.

Table 2 lists another ringing and recovery category of recoveries up to 200 km: road death figures in this class were lower than those produced using only recoveries inside the study area but higher than those from this research. Comparisons using this method are not strictly valid because deaths also resulted from causes outside the study area. Nevertheless it is shown in Table 3 because many

owls investigators are still working with this technique.

Table 3 shows similar percentages in the death-cause classes for the four owl species. The only deviations from the general pattern were the higher percentages of deaths from natural causes in the Little and Long-eared Owls. The majority of deaths (54–68%) were probably caused by man, as suggested in the first six years of the study (Illner 1981) but these figures were probably exaggerated owing to methodological bias (see above). There was no significant change in the proportion of deaths in each category between the two year periods, 1974–1979 and 1980–1986.

Many publications report higher percentages of owl road deaths after 1970, especially for the Barn Owl (examples in Glutz & Bauer 1980) but this may have resulted from the method which was used (ringing and recovery). Weir (1971), using a similar method to the current one, found about 30% of 39 owls (mainly Tawny Owls) on or by roads during a study between 1964 and 1969, about the same overall mean figure found in this study.

Table 4 shows that there was a highly significant difference in owl-kill rates (deaths per 100 km per year) on roads of different types: on roads with car speeds regularly greater than 80 km/h about 21 times as many owls were killed by cars as on the other roads. Similar differences were found for each of the two species (Barn Owl and Little Owl) for which sufficient data were available to analyse separately. Furthermore, trunk roads with highest average car speeds produce highest figures of killed Barn Owls. This difference can only partly be explained by a higher reporting rate (through finds by the road maintenance department; subtracting these finds, the difference remains significant: $p < 0.05$, χ^2 test).

Altogether 1.6 road deaths per 100 km a year were found (Table 4). Two road surveys with enough material on owls but with greater searching activities can be used for comparisons: Bourquin (1983) found 80 owls between 1964 and 1981 on a trunk road in Switzerland (12.4 per 100 km per year, mainly Barn

Table 3. Causes of owl death or fatal disablement in the study area, 1974–1986 (birds older than three months only).

Cause of death	Barn	Little	Long-eared	Tawny	All Owls
Found dead on road	30%	33%	25%	28%	30%
Other deaths caused by man*	38%	26%	30%	37%	33%
Natural deaths**	12%	24%	40%	18%	20%
Unknown	20%	18%	5%	18%	17%
Total finds	74	46	20	11	151

*Includes: found on railway, under powerlines, in chimneys or vertical pipes, entangled in barbed wire, drowned in water tank, shot, poisoned and killed by domestic cat or dog.
**Includes: starved, killed by wild animals, parasites or natural disease with no sign of starvation–disease–human influence.

Table 4. Numbers of owls found dead on roads in relation to road type, 1972–1988 (1972/73 and 1987/88 together seven finds).

Road type	Barn	Little	Long-eared	Tawny	All owls
Car speed >80 km/h					
–Trunk road (16)	4.8 (13)	0.4 (1)	0.7 (2)	0	5.9 (16**)
–Other road outside village (70)	1.1 (13)	0.3 (4)	0.2 (2)	0.1 (1)	1.7 (20)
–Road inside village (18)	0.5 (1)	2.0 (6)	0.3 (1)	0.3 (1)	2.9 (9)
Total >80 km/h (104)	1.5 (27)	0.6 (11)	0.3 (5)	0.1 (2)	2.5 (45)
Car speed <80 km/h (100*)	0	0.1 (2)	0	0	0.1 (2)
No exact location of road casuality	(0)	(3)	(1)	(1)	(5)
Total	0.7 (27)	0.5 (16)	0.2 (6)	0.1 (3)	1.6 (52)

Values given in the table are the number of owls found dead per 100 km of road and per year. The total length of road in each category is given in brackets after the category name, the total number of finds is given in brackets after each value.
χ^2 tests showed there to be significantly more owls killed on roads with cars faster than 80 km/h than those with slower traffic ($p < 0.001$), for both species for which there was an adequate sample; Barn Owl ($p < 0.001$) and Little Owl ($p < 0.05$). Additionally Barn Owl showed a significant difference in death rates when comparing trunk roads and other roads outside villages ($p < 0.01$).
*Minimum figure, only major roads were included in the only town and larger villages.
**Finds from the road maintenance department: Barn Owl, 5; Little Owl, 1; Long-eared Owl, 2.

Owls), and 14 owls between 1960–61 on British roads (2.4 per 100 km per year; mainly Tawny Owls) found by Hodson & Snow (1965).

Vehicle speed has been suggested as an important factor influencing the mortality of birds on roads in several studies (Martens 1962, cited in Bergmann 1974; Hodson 1960; Hansen 1969; Lidauer 1983; Smettan 1988; Shawyer 1987). However none of these studies have included data quantifying traffic speed and separating it from traffic volume. Road death rates appeared to be little affected by the density of traffic. Speed of the vehicles seemed to be more important. My observations suggested that high kill rates occurred on roads or road sections with high car speeds and high owl breeding populations, independent of the degree of traffic density (from moderate to heavy). Bergmann (1974) even established a negative correlation of bird kill rates with traffic density.

In the case of Little Owl the road death rate was increased if the road passed through a village whilst the opposite tendency was apparent for the Barn Owl (Table 4). Comparing the frequencies inside and outside villages for the two species gave a significant difference ($p < 0.01$, Fisher's exact test). This is plausible because most Little Owls breed in villages (Illner in prep.) and their home ranges are small in relation to the Barn Owl (Glutz & Bauer 1980 and personal observation).

It is not certain that all owls killed on fast roads actually collided with a car travelling at a speed greater than 80 km/h but the large difference (Table 4) makes it probable. Furthermore, the author experienced four instances (three Barn Owls and one

Little Owl) of near collisions (1–2 m from the car's windscreen) at speeds of 60–80 km/h. A car speed of about 80 km/h seems to be a critical limit for most owls, perhaps 60–80 km/h for most Little Owls. Testing the frequencies of Little Owl finds on roads with car speeds less than and greater than 80 km/h relative to those of the other three owl species, the difference tended towards statistical significance ($p < 0.1$, Fisher's exact test).

Table 5 presents road deaths in relation to the owl breeding populations, which is a better way to estimate the importance of mortality factors than the use of relative death rates. Absolute minimum percentages of adult road deaths were obtained by subtracting estimated percentages of first year birds (25–50%). These were multiplied by an estimate of the "unknown" factor (see explanation in Table 5) to the improved estimate of the minimum percentages. This last figure was highest for the Barn Own (6.5%) and smallest for the Tawny Owl (1%). A particularly high number of seven Little Owls killed by cars occurred in the breeding season between May and July, corresponding with the results of Dunthorn & Errington (1964). Killing of adult owls in the breeding season could have a particular detrimental effect on population levels (Newton 1979).

Using figures for total annual adult mortality from Glutz & Bauer (1980), it was possible to obtain an estimate of the proportion of adult mortality occurring on roads. The calculated values were about 10–15% for both Barn and Little Owls. Therefore road deaths could have contributed to the observed long-term decline of the two species, although a reduction of food resources appeared to be the major factor (Illner 1988 and in prep.). No evidence was

Table 5. Estimates of road death mortality rates in owl populations, 1974–1986.

	Barn	Little	Long-eared	Tawny	All owls
Mean no. territorial individuals/year (from Illner 1988)	26	79	34	48	187
Mean no. road deaths/year	1.7	1.2	0.4	0.3	3.5
% killed on roads	6.5	1.5	1.0	0.5	1.9
Estimated % of first years in road deaths*	50	25	40	40	40
% of adult road deaths	3.3	1.1	0.6	0.3	1.2
Estimated "unknown" factor**	2.0	3.5	2.5	3.0	2.5
Corrected % of adults killed on roads	6.5	4.0	1.5	1.0	3.5

Annual mortality rates are given.

*The values based on ten ringed Barn and twelve ringed Little Owls. The values for Long-eared and Tawny Owls are about the the mean value of Barn and Little Owls percentages because only one ringed road victim was found.

**The estimation of the "unknown" factor of owls killed by cars but not recorded was based on the assumption that at least half of the corpses could not be found. Heijnis (1980) showed experimentally that 70% of marked dead Starlings (*Sturnus vulgaris*) laid out in the evening under power lines had disappeared next morning, most likely taken by scavenging mammals. The minimum figure of 50% disappearance rate for the Barn Owl (giving an "unknown" factor of two) reflected its greater chance of being found; (a) it has the most conspicuous plumage, (b) it is in greatest demand for stuffed specimens, and (c) lives in more open habitats. The Little Owl is the least conspicuous species (small and cryptic coloured) so one would expect the number of overlooked corpses to be greatest, with the Tawny and Long-eared Owl lying between these two extremes.

found to suggest that populations of Long-eared and Tawny Owls had been seriously affected by collisions with vehicles. These estimates assume, amongst other things, that collisions were the ultimate cause of death and that it was additive to the overall mortality (Newton 1979). The fact that many adult owls killed by cars appeared to be in good condition at least partially confirmed this assumption (see also Sutton 1972, cited in Dunthorn and Errington 1964; Weir 1971). Furthermore it is also possible that some owls colliding with vehicles were only injured and died later by other proximate causes (Finnis 1960; Hodson & Snow 1965) or suffered reduced fitness (Lidauer 1983), underestimating the importance of road accidents.

Using a different calculation procedure, Shawyer (1987 and personal communication) estimated the road death mortality rate (first year birds included) of British Barn Owls to be 28–36% of the anticipated annual death rate. Using the author's calculation procedure (Table 5) in place of Shawyer's, a road mortality rate of 2.5–10% from the same basic data is estimated. With data of Schönfeld *et al.* 1977, a mean road death figure of about 2–4% (adults) was calculated for the mean annual Barn Owl breeding

population in East Germany. Data from Weir (1971) using the same calculation procedure (no differentiation between age classes practicable) result in a figure of 1.4% of the owl breeding populations (mainly Tawny Owls). In both cases the estimates of breeding populations are imprecise but they do illustrate the large overestimate of the importance of road deaths obtained from the ringing and recovery method, as found in this study.

Acknowledgements

I am grateful to Chris Husband for improving the English text, and Dr S. Percival and an anonymous referee for constructive comments on the first draft of this paper.

References

Anderson, D.R., Burnham, K.P. & White, G.C. 1985. Problems in estimating age-specific survival rates from recovery data of birds ringed as young. *Journal of Animal Ecology*, 54: 89–98.

Bergmann, H.H. 1975. Zur Phänologie and Ökologie des Strapentods der Vögel. *Die Vogelwelt*, 95: 1–21.

Bezzel, E. 1982. *Vögel in der Kulturlandschaft*. Ulmer, Stuttgart.

Bourquin, J.D. 1983. Mortalite des rapaces le long de l'autoroute Geneve-Lausanne. *Nos Oiseaux, 37*: 149–169.

Bunn, D.S., Warburton, A.B. & Wilson, R.D.S. 1982. *The Barn Owl*. Calton, Poyser.

Cavé, A.J. 1977. Pitfalls in the estimation of age-dependent survival rates of birds from ringing and recovery data. *Die Vogelwarte, 29 (Sonderheft)*: 160–171.

Dunthorn, A.A. & Errington, F.P. 1964. Casualties among birds along a selected road in Wiltshire. *Bird Study, 11*: 168–182.

Finnis, R.G. 1960. Road casualties among birds. *Bird Study, 7*: 21–32.

Glue, D.E. 1971. Ringing recovery circumstances of small birds of prey. *Bird Study, 18*: 137–146.

Glutz von Blotzheim, U.N. & Bauer, K.M. 1980. *Handbuch der Vögel Mitteleuropas, Vol. IX*. Akademische Verlagsgesellschaft, Wiesbaden.

Hansen, L. 1969. Trafikdoden i den Danske dyreverden. *Dansk Ornitologisk Forenings Tidsskrift, 63*: 81–92.

Heijnis, R. 1980. Vogeltod durch Drahtanfluge bei Hochspannungsleitungen. *Okologie der Vogel (Sonderheft), 2*: 111–129.

Hodson, N.L. 1960. A survey of vertebrate road mortality, 1959. *Bird Study, 7*: 224–231.

Hodson, N.L. & Snow, D.W. 1965. The road deaths enquiry, 1960–61. *Bird Study, 12*: 90–99.

Illner, H. 1981. Populationsentwicklung der Eulen (*Strigiformes*) auf einer Probeflache Mittelwestfalens 1974–1979 und-bestandsbeeinflussende Faktoren, insbesondere anthropogener Art. *Ökologie der Vogel (Sonderheft), 3*: 301–310.

Illner, H. 1988. Langfristiger Rückgang von Schleiereule *Tyto alba*, Waldohreule *Asio otus*, Steinkauz *Athene noctua* and Waldkauz *Strix aluco* in der Agrarlandschaft Mittelwestfalens 1974–1986. *Die Vogelwelt, 109*: 145–151.

Lidauer, R.M. 1983. Knochenfrakturen bei Stadtamseln *Turdus merula*. *Ökologie der Vögel, 5*: 11–126.

Löhrl, H. 1950. Vögel als Verkehrsopfer. *Berichte der Staatlichen Vogelschutzwarte Ludwigsburg, 1949*: 132–136.

Newton, I. 1979. *Population ecology of raptors*. Berkhamsted, Poyser.

Sachs, L. 1974. *Ange wandte Statistik*. Berlin, Springer.

Schönfeld, M., Girbig, G. & Stamm, H. 1977. Beitrage zur Populationsdynamik der Schleiereule, *Tyto alba*. *Hercynia Neue Folge, Leipzig, 14*: 303–351.

Shawyer, C.R. 1987. *The Barn Owl in the British Isles: its past, present and future*. The Hawk Trust, London.

Smettan, C.R. 1988. Wirbeltier und Straßenverkehr-ein ökologischer Beitrag zum Straßentod von Säugern und Vögeln am Beispiel von Ostfildern/Württemberg. *Ornithologische Jahreshefte fur Baden-Württemberg, 4*: 29–55.

Uhlenhaut, K. 1976. Unfalle von Schleiereulen durch Kraftfahrzeuge. *Der Falke, 21*: 56–60.

Weir, D.N. 1971. Mortality of hawks and owls in Speyside. *Bird Study, 18*: 147–154.

Rates of prey delivery to the nest and chick growth patterns of Barn Owls *Tyto alba*

I.K. Langford and I.R. Taylor

Langford, I.K. & Taylor, I.R. 1992. Rates of prey delivery to the nest and chick growth patterns of Barn Owls *Tyto alba*. *In: The ecology and conservation of European owls*, ed. by C.A. Galbraith, I.R. Taylor and S. Percival, 100-104. Peterborough, Joint Nature Conservation Committee. (UK Nature Conservation, No. 5.)

During 1986–88, automatic recording equipment was installed at 19 Barn Owl *Tyto alba* nests in south Scotland during the entire breeding season from laying to fledging.

During laying, prey delivery rate was very high, on average, about 16 items per day. This stabilised to around 8–10 items per day during incubation, increasing from hatching to reach a peak of about 14–16 items per day when the broods were approximately 24–30 days old. Prey delivery rates declined progressively after 30 days but a small number of items were still being delivered 80–90 days after the first chick hatched.

Skeleton growth as assessed by tarsus length was complete by days 35–40. Maximum weights were attained between days 35–40. Wing growth was completed between days 65 and 70.

Average brood size of the nests examined was 3.6. Items brought to the nest were predominantly Field Voles *Microtus agrestis*. Average prey weight during laying and incubation was 26.0 ± 1.7 g and during the chick stage, 25.1 ± 2.4 g. Daily food intake of chicks between 30 and 35 days was estimated to be between 79.6 and 12.8 lg.

I.K. Langford & I.R. Taylor, Institute of Cell, Animal and Population Biology. The University of Edinburgh, Zoology Building, West Mains Road, Edinburgh, EH9 3JU

Introduction

In this paper we present data on the rates of prey delivery to Barn Owl *Tyto alba* nests during the breeding season from laying through to fledging. Data are also presented on the growth patterns of the young.

Previous studies of prey delivery rates have been rather limited; Bussman (1935, 1937) recorded feeding rates at two nests and Ritter & Gorner (1977) at one nest. Growth patterns and rates of growth have been described for *Tyto alba guttata* in Germany (Schonfeld & Girbig 1975) and for Barn Owls in France, in an area of overlap between races *alba* and *guttata* (Baudvin 1975).

Methods

The study was carried out in Dumfriesshire, south Scotland, in predominantly pastoral farmland, interspersed with numerous small, mostly coniferous, woodlands.

Prey delivery rates were studied in 1986, 1987 and 1988 at 19 nests in disused farm buildings. Automatic recording equipment was developed to record visits to the nest. This consisted of an infra-red light source with two infra-red sensors positioned opposite, across the entry hole. The sensors were spaced such that a bird entering interrupted first one sensor then the other. The equipment thus functioned as a direction decider from which a bird entering could readily be distinguished from a bird leaving. The output of the direction decider was in digital form and was recorded directly onto Squirrel data loggers which could be later unloaded and printed via a computer. The direction decider was also able to control a camera and flash to record, photographically, prey deliveries to the nest.

Five potential sources of error were identified with the use of the automatic recording equipment to estimate prey delivery rates.

1. Random signals through equipment malfunction.

2. Failure to record every entry/exit.

3. Entries to the nest area by other species.

4. Visits to the nest without prey.

5. Interpretation of read outs.

Each of these was rigorously tested. The possibility of random signals was examined by simultaneous photography at four nests and direct observations at five nests over a large sample period. Recordings were made only at nest sites where there was a single entrance so that birds always had to pass the sensors. At all, a specially constructed tunnel was installed such that entry or exit without breaking the infra-red beam was impossible. The equipment was tested thoroughly in the laboratory to ensure that every time the beam was broken the event was registered on the data loggers and it was subsequently tested at intervals of about five days in the field. The comparison of direct observations and automatic recording (above) also served as a test.

Errors arising from other species entering the nest area were examined using the large sample of photographs taken at the nest (above). The photographic record was also used to determine the percentage of visits made without prey. All photographic recording was undertaken over 24 hrs and thus was better than direct observations that could only be done in daylight.

The main problem presented by the interpretation of the recordings arose from birds hesitating at entry and the female leaving during incubation and brooding. With the former, multiple recordings occurred in very quick succession, within less than a minute, as the bird repeatedly broke the beams. Such recordings were treated as single events. The exit of an incubating female was a clearly recorded event but there was the possibility of confusion between her return and a feeding visit by the male. Visits by the male were of very short duration such that an entry was quickly followed by an exit. A returning female was characterised by an entry without a corresponding rapid exit record. This pattern was confirmed by direct observation.

Prey delivery rates were quantified as the number of items per 24 hrs, from 1200 hrs one day to 1200 hrs the next.

Chick growth was investigated during 1980–84, at 158 nests involving 572 chicks. Each nest was visited approximately once every 10 days but as visits to different nests were at slightly different stages of growth, mean daily growth rates from within the sample could be examined. Measurements were made of tarsus length (to nearest 1.0 mm), wing length, taken as the maximum, flattened, straightened chord (to nearest 1.0 mm) and weight to nearest 1.0 g up to day 10 and to nearest 5.0 g thereafter).
Information on the weights of items delivered to the nests was obtained by visiting nests in the morning when prey items remained uneaten. These could then be weighed and values of prey weight established.

Results

Validation of the data

Random signals would have been evident as blank photographs when both automatic camera and data loggers were simultaneously connected to the infra-red equipment. Of 632 photographs taken during the testing, none were blank. All events registered on the data loggers were visits by real objects. During the testing by direct observation, 163 ins were observed and 163 recorded and 181 outs were observed with the same number recorded on the logger. Of the 632 events recorded by photography, only two (0.16%) were not Barn Owls (one moth and one Pied Wagtail). Of 630 visits by Barn Owls 614 (97.2%) were visits with prey. From direct observation, 162 of the 163 arrivals seen were owls carrying prey. On the basis of these tests it is concluded that the automatic equipment provided a reliable recording of feeding visits to the nest.

Chick growth

On the day of hatching, mean tarsus length was 9.0 mm. Maximum growth rates occurred around day 15 and maximum length was attained between 35 and 40 days from hatching (Figure 1).

Wing length averaged 13.0 mm on the day of hatching. Primary quills were first evident at 13–14 days and maximum wing length was reached between 65 and 70 days. Between 25 and 60 days wing growth rate averaged 5.1 mm per day (Figure 2).

Weights at hatching averaged 13–14 g. Maximum weights of just under 400 g on average, (with some individuals attaining up to 450 g) were attained between 35 to 50 days. Thereafter weights declined slightly to an average of 340–350 g before fledging.

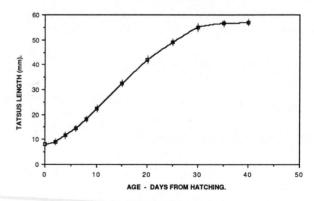

Figure 1. Pattern of tarsus growth of Barn Owl. To day ten, values are means at two day intervals, from ten days on, means at five day intervals. 95% confidence intervals are shown.

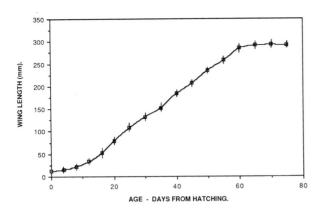

Figure 2. Pattern of wing growth of Barn Owl. To day 20 values are means at four day intervals and from day 20 onwards, means at five days intervals.

Maximum rates of weight gain occurred between days 5 and 20 (Figure 3). Weight was the most variable of the growth parameters measured being influenced by the amount of food in the gut and also varying in relation to brood size and prey abundance (Taylor in prep.).

Prey delivery rates

During laying prey delivery rates were high, averaging 16.4 items per 24 hrs (1200 hrs to 1200 hrs). This fell sharply to a level of 8–11 items per 24 hrs during incubation (Figures 4 and 5). Levels increased from hatching to reach a peak of around 14–16 items/24 hrs approximately 30 days after hatching of the first chick. Thereafter, there was a steady decrease but a low level of prey delivery was still recorded about 90 days after the first egg hatch.

Mean brood size of the nests studied was 3.6 so that at maximum feeding rates each received, on average, about 4 items/24 hrs. Mean prey weight was 26.0 ± 1.7 g during laying and incubation and 25.1 ± 2.4 g during the chick stage.

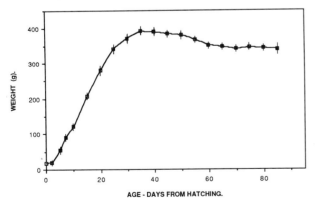

Figure 3. Pattern of weight change in Barn Owl chicks. Values are means at two day intervals to ten days, at five day intervals thereafter.

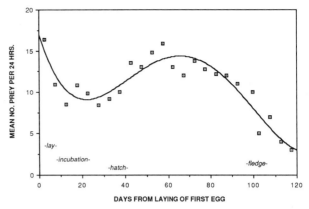

Figure 4. Rates of prey delivery to 19 Barn Owl nests. Values are means of five day intervals. Mean brood size was 3.6.

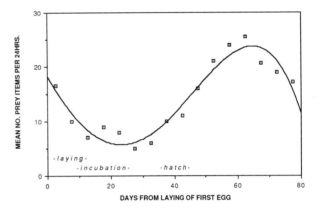

Figure 5. Rates of prey delivery to a brood of six Barn Owls. Values are means of five day intervals.

Discussion

The very high rate of prey delivery during the laying period was common to all nests examined. The amounts involved were around 16 items per day almost all of which were Field Voles. Thus the weight of prey delivered, on average 436 g per 24 hrs, would seem to be far in excess of the females' requirements. At most nests considerable quantities of surplus prey were evident during laying with lower quantities through incubation, rising again during the first three weeks of chick growth, a phenomenon previously observed by Baudvin (1980) in his French population. Caching of apparently surplus amounts of food may simply be an insurance against any possibility of food shortage at critical times as might be caused by adverse weather conditions prohibiting hunting or seriously reducing success. Similar caching behaviour has been reported in other species, e.g. Tengmalm's Owl *Aegolius funereus* (Korpimaki 1987).

There may also be an additional explanation for such high delivery rates during laying in Barn Owls. From before laying until the clutch is complete, male Barn

Owls roost at the nest site with the female and minimise the time spent away from the female, probably exhibiting mate guarding behaviour (Taylor in prep.). The presence of a food cache might also reduce the chances of the male having to spend long periods away from the nest should hunting conditions deteriorate by providing a food source for himself as well as the female.

Smaller amounts of prey are sometimes found at the nest when the chicks are older than 20 days (Baudvin 1980) but by about 30 days it is rare to find uneaten prey (pers. obs.). Thus from then onwards recorded prey delivery rates were probably a good measure of daily food intake by the chicks. From 50 days onwards there was a complication in that some of the chicks at the nests studied could climb outside the automatic recorder to be fed. Thus although feeding rates declined from when the oldest chick was 30 days, recorded rates after 50 days were less than actual feeding rates. From 30 to 50 days recorded rates of prey delivery were an accurate measure of food intake rates. Thus at the maximum rate from 30–35 days, 14–16 items were eaten or 3.9–4.3 items per chick, at a mean weight of $25.1 \text{ g} \pm 2.4 \text{ g}$ per item giving a total daily intake estimate of 79.6 and 128.1 g.

Skeleton growth as assessed by tarsus length was completed between 35 and 40 days. Thus it appears that food delivery rates begin to decline once skeleton growth is complete. Chick weights also started to decline from about 35–40 days onwards. Wing growth was completed by about day 70 but prey was still being delivered to the nest when the youngest chicks were about 85 days old. Thus it is evident that there is a period of post-fledging dependence on the parents of at least 15 days.

Acknowledgements

We thank Buccleuch Estates and many farmers for permission to study Barn Owls on their properties. Special thanks are due to Dr Andrew Sandford for giving the benefit of his considerable talents in helping to develop the automatic recording machinery and to Tom Irving who helped in many ways. A. Langford, M. Osborne and F. Slack helped with field work. We thank the Nature Conservancy Council, the World Wide Fund for Nature and the British Ecological Society who provided financial assistance.

References

Baudvin, H. 1975. Biologie de reproduction de la chouette effraie *Tyto alba* en Cote D'Or. Premier resultats. *Le Jean Le Blanc*, *14*: 1–51.

Baudvin, H. 1980. Les surplus de proies au site de nid chez la chouette effraie, *Tyto alba*. *Nos Oiseaux*, *35*: 232–238.

Bussmann, J. 1935. Der Terragraph am Schleiereulenhorst. *Ornithologische Beobachter, 32*: 175–179.

Bussmann, J. 1937. Biologische Beobachtange die Entwicklung der Schleiereale Schweiz. *Arch. fur Orn. 1*: 377–390.

Korpimäki, E. 1987. Prey caching of breeding Tengmalm's owls, *Aegolius funereus*, as a buffer against temporary food shortage. *Ibis, 129*: 499–510.

Riter, F. & Gorner, M. 1977. Untersuchungen uber die Beziehung zwischen Futterungsakivitat und Beutetierzahl bei der Schleiereule. *Der Falke*, *24*: 344–348.

Schonfeld, M. & Girbig, G. 1975. Beitrage zur Brutbiologie der Schleiereule, *Tyto alba* unter besonderer Berucksichtigung der Abhangigkeit von der Feldmausdichte. *Hercynia, 12*: 257–319.

The dynamics of depleted and introduced farmland barn owl, *Tyto alba* populations: a modelling approach

I.R. Taylor and J. Massheder

Taylor, I.R. and Massheder, J. 1992. The dynamics of depleted and introduced farmland barn owl, *Tyto alba* populations: a modelling approach. *In: The ecology and conservation of European owls*, ed. by C.A. Galbraith, I.R. Taylor and S. Percival, 105-110. Peterborough, Joint Nature Conservation Committee. (UK Nature Conservation No. 5.)

A deterministic population model incorporating population parameter values derived from a long term field study was used to simulate the dynamics of depleted and introduced populations of barn owls under different conditions of prey density. At high prey densities barn owl populations are capable of rapid growth, doubling in 3 to 4 years. Growth occurred only where mean population productivity exceeded 3.2 young per pair. The significance of captive breeding and release schemes is discussed.

I.R. Taylor & J. Massheder, Institute of Cell, Animal and Population Biology, University of Edinburgh, Zoology Building, West Mains Road, Edinburgh EH9 3JT.

Introduction

The Barn Owl, *Tyto alba*, has undergone long term declines in most of Europe and in many parts of North America (Blaker 1933; Prestt 1965, Braaksma 1980, Stewart 1980, Ziesemer 1980, Bunn *et al.* 1982, Pikula *et al.* 1984, Colvin 1985, Shawyer 1987, Marti 1988). In Britain an accelerated decline was recorded in the late 1950s and early 1960s correlating with major changes in farming practice, involving habitat change and use of pesticides (Prestt 1965). A similar decline was described from Ohio which Colvin (1985) was able to attribute to a reduction in the quality of farmland habitat.

A number of measures have been discussed to reverse these trends including habitat modification to increase prey density (Colvin *et al.* 1984, Taylor 1989) and the release of captive bred birds (Shawyer 1987).

In this study we incorporate life-history parameters for Barn Owls, as derived from a long-term study in farmland habitat (see below), into a population simulation model. The model is simplified, the objective being to gain an understanding of likely trends and rates of change of barn owl populations under different conditions of prey density. In the field study prey density varied from year to year, providing a natural experiment and enabling population parameters to be quantified at different prey densities.

Population simulation models have often been used to predict rates of recovery of depleted populations and rates of increase of introduced populations. They have therefore been useful tools in conservation planning. Specific cases have included species recovering after severe habitat loss (e.g. Whooping Crane, *Grus americana*, Blinkley & Miller 1988), overhunting (e.g. Australian Freshwater Crocodile, *Crocodylus johnstoni*, Smith & Webb 1985) and the effects of pesticides (e.g. Osprey *Pandion haliaetus*, Henny & Wight 1969 and Peregrine Falcon *Falco peregrinus*, Barclay & Cade 1983, Lindberg 1983, Grier & Barclay 1988).

Methods

Model description

The model, described diagrammatically in Figure 1, was deterministic and although less realistic than a stochastic model gave a clearer picture of the trends in population under different conditions. It assumed that nest sites were not limiting and did not

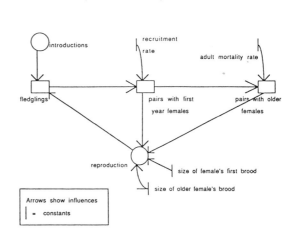

Figure 1. Diagram of population model structure.

incorporate density dependent functions. In the field study, no evidence of density dependent productivity or mortality of nest site limitation was found. The model was therefore valid only for investigating the behaviour of depleted, recovering or introduced populations below the level at which density limitation might be expected to occur.

The model population was subdivided into 3 components; first year birds before their first breeding season, pairs with females in their first breeding season and pairs with older females. The model was operated one year at a time. Rates of recruitment, from the field study, were applied to the first year component to give the number entering the breeding population. All birds were assumed to start breeding when one year old, i.e. in their second calendar year of life. In the field study only 7% trapped were breeding for the first time as two year olds. In the model, mortality of age classes greater than one year was applied as a single event immediately prior to breeding. All surviving birds were assumed to breed.

When considering introductions, the model assumed that introduced birds had the same recruitment and survival rates and the same breeding performance as wild birds. It assumed that no deleterious genetic effects such as inbreeding depression resulted from captive breeding or outbreeding depression from interbreeding of individuals from distant populations. Introduced birds were added at the newly fledged stage. No simulations were done with birds released as breeding age adults.

Values of parameters used in the model

The values of all parameters were the empirical results of an 11 year field study (1979–1989, see below), the results of which will be presented in detail elsewhere. During the study, mortality rates, recruitment rates, breeding performance and movements were quantified for the whole population under investigation. An annual index of field vole *Microtus agrestis*, abundance, the owls' main prey species, was determined each year by trapping. All population parameters were significantly linearly correlated with variations in prey density. Breeding performance was significantly lower for first year females than for older females. Dispersal distances were short and immigration and emigration were not significant contributors to population change.

Field study area

The field study area, situated in south Scotland covered 300 km² of farmland at altitudes between 0

and 125 m above sea level. Farming was predominantly pastoral involving dairy and sheep production with field habitats mainly of rotational improved pasture (40% of total area), hay and silage (together, 18%). Cereal crops occupied only 11% of the area. The farmland was interspersed with numerous small coniferous and deciduous woodlands (9% of area). Barn Owls utilized edge habitat especially woodland edges for foraging. There was little application of pesticides either historically or during the study. The climate was similar to that of much of lowland Britain with average January temperatures between 3.0 and 3.5°C and snow cover of 10–20 days per annum. Rainfall, at 80–100 cm per annum was somewhat above the average range of 60–80 cm for lowland areas (Metereological Office 1975a, b, 1977, Chandler & Gregory 1976).

Testing the model's predictions

In 1982 there were 20 pairs of Barn Owls breeding in the study area. Between then and 1986 the population underwent a short-term decline. Following 1986 the population recovered towards the 1982 level. The decline period was used to test the model's ability to predict population trends. The model was programmed with the mean values of productivity, recruitment and survival for 1982 to 1986, from which estimates were generated of the population size each year to 1986. Predicted population sizes and the trend were compared with the known population sizes and trend.

Results

Using the model, the behaviour of depleted wild populations and introduced populations was examined under constant conditions of prey density, over a period of 20 years. The model was run with appropriate population parameter values for 4 prey densities representing the lowest and highest encountered in the field study and the 2 equidistant densities between (Table 1).

Natural populations

Simulated natural populations were initiated at a level of 50 pairs. Population growth was closely related to prey density (Figure 2). At the highest prey density, growth over 20 years followed the exponential relationship, $y = 48.5 \times 10(0.09x)$, $r = 1.0$; where $y =$ number of pairs and $x =$ years. Age structure was stable with 31% first year breeders (Figure 3). Mean population productivity was also

Table 1. Values of population parameters used in this model at different levels of prey density. Levels 1 to 4 correspond to trapping indices of 2, 17, 19 and 45. The index is the mean number of voles trapped over 5 consecutive days at 6 sample sites each with 48 traps.

Prey density level	Recruitment rate	Adult mortality rate	Reproduction rate of 1 year old females	Reproduction rate of >1 year old females
1	0.05	0.29	1.60	2.40
2	0.10	0.24	2.00	2.90
3	0.15	0.20	2.40	3.40
4	0.21	0.15	2.70	4.00

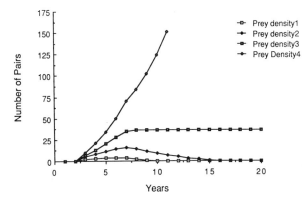

Figure 2. Numerical response of depleted natural populations to differentlevels of prey density. Density levels as in Table 1.

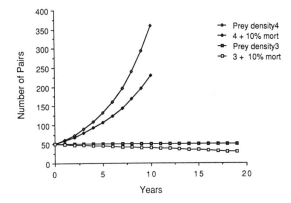

Figure 4. Effects of increasing mortality rates by 10% on the growth of simulated populations at prey levels 3 and 4 (see Table 1).

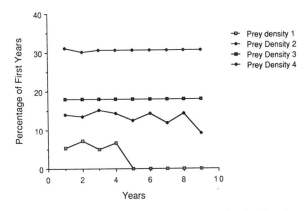

Figure 3. Percentage of first year breeders in simulated natural populations at different levels of prey density. Density levels as in Table 1.

A further simulation was done to examine the effects of a 10% increase in mortality rate (and similar reduction in recruitment rate), such as could result from pesticide poisoning (Newton *et al.* 1991). The model was run at the two highest prey densities. At the higher of these, the potential growth of the population was considerably reduced. Without the added mortality, the population reached 360 pairs in 10 years but, with the 10% extra mortality, the increase was to only 229 pairs (62%). At the lower prey density, zero growth occurred without the added mortality but, with the extra mortality, the population declined to 60% of the original level after 20 years (Figure 4).

Introduced populations

Simulations of introduced populations were done at the four prey densities by adding 100 birds, with sex ratio of unity, at the fledging stage in each of five years. The model was run for 20 years in each case.

At the highest prey density, the introductions resulted in an established population which, after six years, increased exponentially (Figure 5). The percentage of first year breeders declined from 100% in year 2 to stabilise at 31% from year 7 (Figure 6). As a

stable at 3.6 young per pair. At the next lower prey density, the population showed zero growth, remaining at 50 pairs. Age structure was stable at 18% first year breeders and mean productivity was 3.2 young per pair. At the two lower prey densities, population growth was negative and extinction occurred after 15 and 9 years. In the former case, productivity was approximately 2.8 young per pair and first year breeders comprised 10–14% of the population. In the lowest prey density treatment, productivity was 2.3 young per pair and only 5–7% of the population were first year breeders.

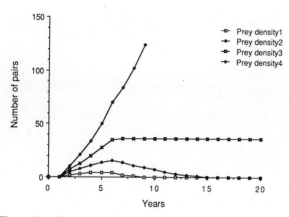

Figure 5. Numerical effects of simulated introductions at different levels of prey density (see Table 1). In each case, 100 birds were introduced in each of 5 years.

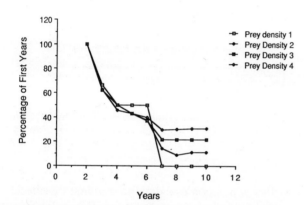

Figure 6. Percentage of first year breeders in simulated introduced populations at different levels of prey density (see Table 1).

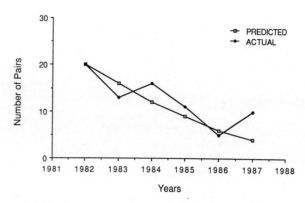

Figure 7. A comparison of the real population trend observed in the field study area with the trend predicted by the population simulation model from 1982 to 1986.

Testing the model's predictions

The model predicted the number of pairs in the field study population to decline steadily from 1982 to reach a level of four pairs in 1986. This was very close to the changes in the real population; a downward trend with five pairs in 1986. The model thus accurately predicted the real trend (Figure 7).

Discussion

The results of this modelling exercise must be interpreted with care. It is particularly important to emphasise that the objective was to examine the response of depleted and introduced populations below the levels at which competition for resources, such as nest sites, might occur. Thus the simulations are pertinent, for example, to the initial response of depleted populations which experience large scale improvements in habitat quality. As populations grow towards levels at which resources become limiting and density dependence occurs, their numerical response, expressed as the number of breeding pairs, will be expected to deviate considerably from that predicted by this simple model.

The deterministic nature of the model meant that it was unable to simulate short-term fluctuations of the real population but, given this limitation it was nevertheless able to simulate, with good accuracy, the overall trends in the field study population.

The wider applicability of the results generated by the simulations depends on how representative the study area was of the farmland habitat used by Barn Owls in Britain and hence the extent to which the values of the parameters used in the model are valid for other areas. Most important is the relationship

consequence of this change in age structure, productivity increased from 2.6 young per pair to 3.6 per pair from year 7 on. At the next lower prey density, a stable population of 26 pairs was established after 6 years. First year breeders declined from 100% in year two to 22% from year 7 onwards. Productivity increased from 2.3 to 3.2 young per pair.

At the two lowest prey densities, the introductions failed to establish self-sustaining populations. At the lowest of these, the release of 500 individuals over five years resulted in a maximum of only four wild breeding pairs and the population become extinct three years after releases ceased. Productivity reached a maximum 2.0 young per pair. The next higher prey density simulation resulted in a maximum population of 16 pairs with a maximum productivity of 2.8 young per pair. Population extinction occurred 9 years after releases ceased.

between productivity, recruitment and mortality rates and factors that alter the values of these parameters relative to each other will affect the general validity of the results. Deviations in mortality rates (and hence also recruitment rates) will have the greatest effect. Factors considered to influence mortality rates, apart from prey abundance, are weather conditions, particularly in winter, pesticides such as dieldrin and second generation rodenticides and road traffic. Winter weather conditions in the study area, particularly temperatures and snow duration were typical of much of lowland Britain with the exception of the extreme south west of England (Metereological Office 1975a, 1975b). The predominantly pastoral nature of the study area with little use of pesticides was characteristic of most of western Britain but eastern, cereal growing areas have been subject to greater pesticide use. There is evidence that the use of aldrin/dieldrin increased Barn Owl mortality in these areas in the 1960s and 1970s but, with restrictions on the use of these pesticides, no deaths attributable to them have been recorded in the late 1980s (Newton *et al.* 1991). Present use of second generation rodenticides has also increased mortality. The extent of this has not been quantified but a value of at least 10% is not unlikely in some areas (Newton *et al.* in press).

The effect of road traffic mortality is difficult to assess. Road casualties, on average, weigh significantly less than healthy birds (Taylor in prep) and it is likely that, for a high proportion of them, collision with traffic is only a proximate cause of death with prey abundance and starvation as the ultimate cause. There are no reliable estimates of the extent of losses through collision.

The effect of a relative increase in mortality rates, from whatever source, would be to reduce population growth below that predicted by the model with the parameter values given.

Barn Owl population growth was shown to be highly sensitive to changes in prey density. Given high average prey densities, with no added mortality from factors such as pesticides, depleted populations were capable of rapid increase, doubling in 3 to 4 years. Thus, provided suitable nest sites are available, a conservation strategy which involves increasing the amount of high prey density habitat should be capable of yielding positive results quickly. This should be possible even with a small increase in mortality rates (e.g. through pesticide contamination or road accident).

The critical level of productivity suggested was of the order of 3.2 young per pair. Populations

characterised by this level were capable only of maintaining themselves. To achieve rapid growth, a productivity closer to 3.5 young per pair would probably be needed.

Conservation implications

Quantification of population productivity rates should be a reasonably accurate method for the assessment of the success of conservation measures, provided they are established by rigorously conducted field research which removes sources of bias, such as missing unsuccessful or late breeders, overestimating success by not visiting nests immediately before fledging, or recording only consistently or frequently used sites. Mean values over five or six years would be needed to take account of short-term variations caused by fluctuations in prey density around mean levels. Population age structure could also be a useful indicator, although clearly more difficult to estimate than productivity.

The precise manner in which improvements to habitat are done is important and is probably best tackled locally as well as nationally. In the first instance it is probably more appropriate to concentrate on and immediately around areas which still hold relatively high Barn Owl densities. Before an increase in numbers is possible, either by infilling or by peripheral extension, it may be necessary to improve habitat quality within the ranges of existing pairs, such that the population rises above the productivity and survival rates needed for maintenance. Improvement of habitat in adjacent areas alone would probably result in considerably slower recovery rates. Also, the exact nature of habitat change is critical. Vague prescriptions that bring about some improvement in quality but fail to reach the level needed for positive population growth will be of little value. Indeed, they may be detrimental by undermining the credibility of the conservation effort. In-depth ecological research is necessary to establish the nature and amount of habitat needed. Such knowledge can also be used to identify existing patches of high quality habitat. Local population productivity may be increased at least partly by ensuring that nest sites are available in all such patches.

When considering introduced populations the model assumed that such birds survived and bred as well as wild birds under the same conditions of prey density. This seems unlikely and the model's predictions were probably optimistic, but in the absence of appropriate field data it is not possible to make modifications other than to be aware that the numerical response is likely to be lower than that predicted. Nevertheless a number of useful points are illustrated by the simulations. It is clear that introductions will not succeed where habitat conditions with reduced prey densities, that led to the decline in the first place persist. This has been well illustrated in the field in several failed restocking attempts in USA (Marti 1988). Habitat improvements to raise prey density would be needed for introductions to have any chance of success. Under such conditions a depleted natural population could also increase and the validity of releasing captive bred birds must be questioned. At present, the genetic and other population implications of captive breeding and release are unknown. Given this, and the fact that introductions could be carried out at any time in the future, it seems highly prudent to concentrate on habitat improvement and allow natural populations the opportunity to increase first, before considering other courses of action. Obviously very small, isolated remnant populations would be more vulnerable to chance events and may be more prone to extinction. However, in a properly planned long term conservation effort, such areas could be recolonised naturally as other populations expand. Other considerations apart, the introduction of captive bred birds into an area where habitat has been improved and where there is a depleted natural population, would render the interpretation of subsequent population changes considerably more difficult. Even where all released birds were individually ringed, a detailed field study in which the identity of individual breeders, population age structure and productivity were determined, would be needed to identify effects brought about by the habitat changes from those of the releases. With the small sample sizes likely to be involved, it is probable that the two could not be separated and this could considerably delay the attainment of satisfactory conservation measures.

The establishment of captive-bred birds as breeders in the wild is not on its own an adequate criterion by which to assess the success of introduction schemes. As suggested by the simulations, some pairs may become established but following cessation of releases, fail to maintain a stable population.

Acknowledgements

We are grateful to Buccleuch Estates and numerous farmers for permission to study Barn Owls on their land and to T. Irving, I. Langford, P. Bell, F. Slack, S. Abbott and M. Osborne for assistance in the field. Anne Aitken and Connie Fox typed the manuscript.

Funding was provided by the Natural Environment Research Council, Nature Conservancy Council, World Wide Fund for Nature and the University of Edinburgh.

References

Barclay, J.H. & Cade, T.J. 1983. Restoration of the Peregrine Falcon in the eastern United States. *Bird Conservation 1*: 3–37.

Binkley, C.S. & Miller, R.S. 1988. Recovery of the Whooping Crane *Grus americana. Biol. Conserv. 45*: 1–20.

Blaker, G.B. 1933. The Barn Owl in England-Results of the Census. *Bird Notes and News* 15: 169–172.

Braaksma, S. 1980. Gegevens over de achteruitgang van de kerkuil (*Tyto alba guttata*-Brehm) in West-Europa. *Wielwaal. 46*: 421–428.

Bunn, D.S. Warburton, A.B. & Wilson, R.D.S. 1982. *The Barn Owl.* Calton: Poyser.

Colvin, B.A. 1985. Common Barn Owl population decline in Ohio and the relationship to agricultural trends. *J. Field Ornithology, 56*: 224–235.

Colvin, B.A., Hegdal, P.L. & Jackson, W.J. 1984. A comprehensive approach to research and management of common Barn Owl populations. Pp 270–282 in W McComb (ed), *Proc. Workshop on Management of non-game species and Ecological Communities.* Univ of Kentucky, Lexington.

Chandler, T.J. & Gregory, S. 1976. *The Climate of the British Isles.* Longman: London.

Grier, J.W. & Barclay, J.H. 1988. Dynamics of Founder Populations Established by reintroductions. Pp 689–700 in T.J. Cade, J.H. Enderson, C.G. Thelander and C.M. White (eds). *Peregrine Falcon Populations – their Management and Recovery.* The Peregrine Fund Inc Boise, Idaho.

Henry, C.J. & Wight, H.M. 1969. An endangered Osprey population: estimates of mortality and production. *Auk*, 86: 188–198.

Lindberg, P. 1983. Captive breeding and a programme for the reintroduction of Peregrine Falcon (*Falco peregrinus*) in Fennoscandia. *Proc. Third Nordic Congr. Ornithol. 1981.* pp 65–78.

Marti, C.D.L. 1988. *The Common Barn Owl.* Pp 535–550 in W.J. Chandler (ed) Audubon Wildlife Report 1988/89. Academic Press: California.

Meteorological Office. 1975(a). *Maps of mean and extreme temperature over the United Kingdom 1941–1970.* Bracknell: Meteorological Office. (Climatological Memorandum No 73).

Meteorological Office. 1975(b). *Maps of mean number of days of snow cover over the United Kingdom 1941–1970.* Bracknell: Meteorological Office. (Climatological Memorandum No 74).

Meteorological Office. 1977. *Map of Average Rainfall-International Standard Period 1941–1970.* Bracknell: Meteorological Office.

Newton, I., Wyllie, I. & Asher, A. 1991. Mortality causes in British Barn Owls *Tyto alba* with a discussion of dieldrin poisoning. *Ibis,* in press.

Newton, I., Wyllie, I. & Freestone, P. In press. Rodenticides in British Barn Owls. *Environmental Pollution.*

Pikula, J., Beklova, M. & Kubik, V. 1984. The breeding bionomy of *Tyto alba.* Prirodoved Pr Ustava. *Cesk Akad Ved Brne.* 18: 1–56.

Shawyer, C.R. 1987. *The Barn Owl in the British Isles: its past present and future.* The Hawk Trust, London.

Smith, A.M.A. & Webb, G.J.W. 1985. *Crocodylus johnstoni* in the McKinlay River Area. NT VII: A Population Simulation Model. *Australian Wildlife Research,* 12: 541–554.

Stewart, P.A. 1980. Population trends of barn owls in North America. *American Birds,* 34: 698–700.

Taylor, I.R. 1989. *The Barn Owl.* Shire Publications: Princes Risborough.

Ziesemer, F. 1980. Verbreitung, Siedlungsdichte und Bestandsentwicklung der Schleiereule *Tyto alba* in Schleswig-Holstein. *Corax,* 8: 107–130.